AVERTING A LATIN AMERICAN NUCLEAR ARMS RACE

Also by Paul L. Leventhal

NUCLEAR TERRORISM: Defining the Threat
(*co-editor with Yonah Alexander*)

PREVENTING NUCLEAR TERRORISM
(*co-editor with Yonah Alexander*)

Averting a Latin American Nuclear Arms Race

New Prospects and Challenges for Argentine-Brazil Nuclear Co-operation

Edited by

Paul L. Leventhal

President, Nuclear Control Institute, Washington, DC

and

Sharon Tanzer

Executive Director, Nuclear Control Institute, Washington, DC

Palgrave Macmillan

in association with the
NUCLEAR CONTROL INSTITUTE
Washington, DC

ISBN 978-1-349-12101-4   ISBN 978-1-349-12099-4 (eBook)
DOI 10.1007/978-1-349-12099-4

First published in the United States of America in 1992

ISBN 978-0-312-07277-3

Library of Congress Cataloging-in-Publication Data
Averting a Latin American nuclear arms race : new prospects and
challenges for Argentine-Brazil nuclear co-operation / edited by
Paul L. Leventhal and Sharon Tanzer.
     p.   cm.
Proceedings from the Conference on Latin-American Nuclear Co-
operation held in Montevideo, Uruguay, October 11–13, 1989.
Includes index.
ISBN 978-0-312-07277-3
1. Nuclear arms control—Argentina—Congresses.   2. Nuclear arms
control—Brazil—Congresses.   3. Arms race—Argentina—Congresses.
4. Arms race—Brazil—Congresses.   5. Nuclear nonproliferation—
Congresses.   I. Leventhal, Paul.   II. Tanzer, Sharon, 1941–   .
III. Conference on Latin-American Nuclear Co-operation (1989 :
Montevideo, Uruguay)
JX1974.7.A94   1992
327.1'74'098—dc20   91–31595
                                                          CIP

Contents

Acknowledgements

The conference upon which this book is based – held in Montevideo, Uruguay, on October 11–13, 1989 – was made possible by a grant from the Ford Foundation in support of the Nuclear Control Institute's Project on Latin American Nuclear Cooperation. The Institute gratefully acknowledges this grant, as well as the thoughtful suggestions from Enid Schoettle and Christopher J. Welna of the Ford Foundation. Other grants received by the Institute in support of its Nuclear Oversight Project were also applied to the conference.

Development of the Latin American nuclear conference and this book would not have been possible without the direct assistance and support of a number of organizations and individuals. The Argentine Council on Foreign Relations (CARI) and the Brazilian Institute for the Study of International Relations (IPRI) endorsed the conference and provided immeasurable assistance in developing the agenda and in selecting participants. Ambassador Julio Carasales of Argentina and Minister Gelson Fonseca Jr. of Brazil were of special help in this regard. Dr José Goldemberg, then-rector of São Paulo University, also made a valuable contribution to the planning of the conference, as did Dr John Redick of the University of Virginia.

Daniel Horner of the Nuclear Control Institute and Silvana Rubino of PEITHO worked diligently in Washington and Montevideo, respectively, to ensure that the complex logistics for the conference proceeded smoothly. Whitney Watriss of Washington, D.C., provided invaluable assistance in the preparation of the manuscript. Belinda Holdsworth and Anthony Grahame of Macmillan contributed greatly to the planning and editing of this book. Finally, the unique contribution this conference made to the policies of nuclear cooperation and confidence-building eventually announced by the governments of Argentina and Brazil rests upon the personal commitment and substantive contributions of the conference participants and the authors of papers included in this volume.

Notes on the Contributors

José Bernal Castro is special projects manager of Metalurgica Bellucci S.A. in Buenos Aires. He is the Argentine coordinator of the Committee of Argentine and Brazilian Businessmen in the Nuclear Area (CEABAN) and is secretary of the Argentine Committee of Nuclear Enterprises.

Eduardo Bocco is a researcher in international relations at FLACSO (Facultad Latinoamericana de Ciencias Sociales) in Buenos Aires. His current professional interests include the South Atlantic, military and nuclear cooperation, and the military-strategic dimensions of Argentine-Brazilian cooperation and integration.

Ambassador Julio Carasales, a career member of the Argentine Foreign Service, is the former ambassador to the Conference on Disarmament, to the Organization of American States, and to Denmark. He also served as Under-Secretary for Foreign Affairs and Head of the Foreign Service Institute. He is a professor at the University of Belgrano and the Foreign Service Institute.

Vice Admiral Carlos Castro Madero is the former president of the Argentine National Commission of Atomic Energy and former chairman of the Board of Governors of the International Atomic Energy Agency. He also has served as special adviser to the Director General of the IAEA and as president of an expert advisory group to the Director General. He currently is the general secretary of the National Academy of Sciences of Buenos Aires.

Walter Cibils is president of the Uruguayan National Commission of Atomic Energy and National Director of Nuclear Technology. He is secretary of the National Council of Scientific and Technical Investigations.

Rear Admiral Thomas D. Davies retired from the U.S. Navy to become assistant director of the Arms Control and Disarmament Agency. He headed the Nuclear Non-Proliferation Bureau and chaired the U.S. delegation in treaty negotiations with the Soviet Union on a comprehensive test ban and on environmental warfare. In the navy he was an aeronautical engineer and served as chief of naval development.

Samuel Edlow is president of Edlow Resources Limited. He is the founder and chairman emeritus of Edlow International, a company which trades in uranium and provides consulting services on all aspects of the nuclear fuel cycle.

Miguel Estrada Oyuela is a senior officer in the Department of International Relations in the Argentine National Commission of Atomic Energy.

José Felicio is head of the Commodities Division of the Brazilian Ministry of Foreign Affairs and an alternate to the Governor of Brazil on the Board of Governors of the International Atomic Energy Agency. He also has served as head of the Energy and Mineral Resources Division of the Foreign Affairs Ministry and as a member of the staff of the National Security Council.

Oliveiros Ferreira is an analyst of international relations and politics, specializing in Brazilian military affairs. He is a professor in the department of political science at the University of São Paulo. He is the director of the newspaper 'O Estado de São Paulo' and has served as a consultant to businesses in São Paulo.

Edmundo Fujita is counselor in the United Nations Division of the Brazilian Ministry of Foreign Affairs. He previously served as the first secretary of the Brazilian embassies in Moscow and Tokyo.

Roberto García Moritán is Argentina's Ambassador to the Conference on Disarmament. At the time of the conference he was in charge of Nuclear Affairs and Disarmament in the Ministry of Foreign Affairs.

Victor Gilinsky is an independent consultant on nuclear energy issues. He is a former two-term member of the U.S. Nuclear Regulatory Commission. Before his appointment to the NRC, he was head of the Physical Science Department at RAND and assistant director of policy and programme review for the U.S. Atomic Energy Commission.

José Goldemberg is Secretary of State for Science and Technology of Brazil. At the time of the conference he was the rector of São Paulo University. He is a member of the President's Council on Nuclear Policy and the National Council of Science and Technology. He previously served as president of the Brazilian Society of Physics and the Brazilian Society for the Advancement of Science, and as executive secretary of the Energy Council of the State of São Paulo. He has published numerous technical papers on nuclear physics and other issues and is the co-author of *Energy for a Sustainable World* (1987).

Martin Gomez Bustillo, a member of the Argentine Foreign Service, is an officer in the Department of International Security and Nuclear and Space Affairs in the Argentine Ministry of Foreign Affairs. He holds a bachelor's degree in political science and international relations from the University of El Salvador.

Ambassador Hector Gros Espiell is Foreign Minister of Uruguay. At the time of the conference, he was the president of the Interamerican Court of Human Rights and the Undersecretary for Western Sahara Affairs of the United Nations, as well as the head of the Diplomatic Academy in the Uruguayan Ministry of Foreign Affairs. He is the former Secretary General of the Agency for the Prohibition of Nuclear Weapons in Latin America (OPANAL).

Fernando Henning is construction director of NUCLEN, a company which manages design, construction, and commissioning activities for nuclear power plants in Brazil. He is the Brazilian coordinator of the Committee of Argentine and Brazilian Businessmen in the Nuclear Area (CEABAN).

William Higinbotham is a consultant to the Technical Support Organization of the Department of Nuclear Energy at Brookhaven National Laboratory. Prior to that he was a senior scientist in the Technical Support Organization and head of the Instrumentation Division at Brookhaven. He was the first chairman of the Federation of American Scientists.

Monica Hirst is a senior researcher in international relations at FLASCO (Facultad Latinoamericana de Ciencias Sociales) in Buenos Aires.

Milton Hoenig, a nuclear physicist, is scientific director of the Nuclear Control Institute. He is co-author of Volumes I–III of the *Nuclear Weapons Databook*. In 1979 and 1980 he was in the Non-Proliferation Bureau of the U.S. Arms Control and Disarmament Agency, where he worked on technical and economic issues related to the nuclear fuel cycle.

Helen Hunt is a consultant to the Nuclear Control Institute on security and safeguards issues. She is a member of the Institute of Nuclear Materials Management. She received a BA from Cornell University and an MA from Princeton University in mathematics.

Paul Leventhal is founder and president of the Nuclear Control Institute. He was executive vice-chairman of the Institute's International Task Force on the Prevention of Nuclear Terrorism. He has served as staff director of the Senate Nuclear Regulation Subcommittee and as special counsel to the Senate Government Operations Committee. He was responsible for the investigation, hearings and legislation leading to enactment of the Energy

Reorganization Act of 1974 and the Nuclear Non-Proliferation Act of 1978. He was co-director of the Senate Special Investigation of the Three Mile Island nuclear accident.

Marvin M. Miller is a senior research scientist with the Department of Nuclear Engineering and the Center for International Studies at the Massachusetts Institute of Technology. From 1984 to 1986 he was a Foster Fellow with the Nuclear Weapons and Control Bureau of the U.S. Arms Control and Disarmament Agency and is currently a consultant on proliferation issues for ACDA.

Bernardino Pontes is director of the Department of Education and Research of Brazil's National Commission on Nuclear Energy. He is deputy to the Governor on the Board of Directors of the International Atomic Energy Agency. He previously served as a safeguards official with the IAEA and has held a number of teaching positions in nuclear physics.

Gerardo Quintana is a member of the engineering faculty in the department of physics at the University of Buenos Aires. He is a member of the Academic Council of the Faculty of Engineering and has served in the department of nuclear physics of the Argentine Atomic Energy Commission.

John Redick is an associate professor in the Division of Continuing Education at the University of Virginia. He previously served as research director of the Stanley Foundation and as programme officer and acting director of the W. Alton Jones Foundation. He is the author of *Nuclear Restraint in Latin America: Argentina and Brazil* (1988).

Fernando De Souza Barros is professor of physics at the Federal University of Rio de Janeiro and a member of the Brazilian Physical Society's Commission on Nuclear Questions. He is the former president of the Brazilian Physical Society and is a member of the American Physical Society and a fellow of the Brazilian Academy of Sciences.

Sharon Tanzer is Vice President of the Nuclear Control Institute. She was rapporteur and editor of *The Tritium Factor*, the proceedings of a 1988 arms control workshop co-sponsored by the Institute and the American Academy of Arts and Sciences. She helped to organize the Institute's 1985 Conference on International Terrorism and, the following year, its International Task Force on Prevention of Nuclear Terrorism.

Participants at the Conference on Latin-American Nuclear Cooperation

The following participants attended the Conference on Latin-American Nuclear Cooperation: New Prospects and Challenges, held at Montevideo, Uruguay, 11–13 October 1989.

José Bernal Castro, Argentine Chairman, CEABAN.

Eduardo Bocco, research associate, FLACSO.

Ambassador Julio Carasales, former Argentine Ambassador to the Conference on Disarmament, the OAS and Denmark.

Vice Admiral Carlos Castro Madero, former president, Argentine National Commission of Atomic Energy.

Walter Cibils, president, Uruguayan National Commission of Atomic Energy.

Rear Admiral Thomas D. Davies, former assistant director, U.S. Arms Control and Disarmament Agency.

Samuel Edlow, president, Edlow Resources Limited.

Miguel Estrada Oyuela, Director, International Relations, Argentine National Commission of Atomic Energy.

Fabio Feldmann, deputy, Brazilian Congress; environmental advocate.

José Felicio, Director, Commodities Division, Ministry of Foreign Affairs, Brazil.

Oliveiros Ferreira, professor of political science, University of São Paulo; director, 'O Estado de São Paulo'.

Edmundo Fujita, counsellor, United Nations Division, Brazilian Ministry of Foreign Affairs.

Roberto García Moritán, director, Nuclear Affairs and Disarmament, Argentine Ministry of Foreign Affairs; currently, Ambassador to the Conference on Disarmament.

Victor Gilinsky, nuclear energy consultant, former commissioner, U.S. Nuclear Regulatory Commission.

José Goldemberg, rector, São Paulo University; currently, Secretary of State for Science and Technology, Brazil.

Martin Gomez Bustillo, Foreign Service officer in the Department of International Security, Nuclear and Space Affairs, Argentine Ministry of Foreign Affairs.

Ambassador Hector Gros Espiell, president, Interamerican Court of Human Rights; currently, Foreign Minister of Uruguay.

Fernando Henning, Brazilian co-ordinator, CEABAN.

William Higinbotham, consultant, Brookhaven National Laboratory.

Milton M. Hoenig, scientific director, Nuclear Control Institute.

Helen Hunt, consultant, Nuclear Control Institute.

Paul Leventhal, president, Nuclear Control Institute.

Jose Lorenzo Chain, representative of the Uruguayan Ministry of Foreign Affairs to the National Commission of Atomic Energy.

Marvin Miller, Department of Nuclear Engineering and Center for International Studies, Massachusetts Institute of Technology.

Anselmo Paschoa, associate professor of physics, Pontifical Catholic University, Rio de Janeiro.

Emma Perez Ferreira, former president, Argentine National Commission of Atomic Energy.

Daniel Poneman, author; currently, Director, Defense Policy and Arms Control, U.S. National Security Council.

Bernardino Pontes, director, Department of Education, Brazilian National Commission on Nuclear Energy.

Rear Admiral Oscar Quihillalt, former president, Argentine National Commission of Atomic Energy.

Gerardo Quintana, physics professor, University of Buenos Aires.

John Redick, associate professor, University of Virginia.

Fernando de Souza Barros, professor of physics, University of Rio de Janeiro; former president, Brazilian Physical Society.

Sharon Tanzer, rapporteur, Nuclear Control Institute.

Guillermo Tello Rosas, president, Energy Commission, Argentine Chamber of Deputies.

Introduction

There have been few bright spots in the decades-old, uphill struggle to halt the growth of established nuclear arsenals and to stop the spread of new ones. Nuclear non-proliferation can be a very discouraging business. Thus, it is especially noteworthy when a region troubled by nuclear rivalry – in this case Latin America – beats all the odds and makes a breakthrough toward averting an arms race that most experts regarded as inevitable.

That is what happened on 28 November 1990 when the presidents of Argentina and Brazil met at Foz de Iguacu – the majestic Iguacu Falls marking the spot where the boundaries of their two nations meet. There they signed an accord renouncing the development of nuclear weapons and setting forth a number of no-nonsense approaches to assuring one another that their nuclear establishments are living up to the commitment.

It was a stunning renunciation of military control of the nuclear programmes in both nations. Indeed, it was the first clear sign that the transition from military to civilian rule – which took place in 1983 in Argentina and in 1985 in Brazil – had finally been extended to the nuclear sphere.

Only a year earlier, at a conference held in Montevideo, in neighbouring Uruguay, there was resistance, often passionately expressed, by leading Brazilian and Argentine nuclear figures to a number of proposals for reciprocal inspections, international safeguards arrangements, a test ban, and other measures to avert a regional nuclear arms race. The proposals were put forward by a number of US experts assembled by the conference organizer, the Washington-based Nuclear Control Institute, a private, independent, non-proliferation research centre.

At the same time, there was the scent of change in the air as it became apparent from conference discussions that the two nations now found it in their mutual interest to pursue the difficult process of ending 150 years of mutual suspicion. That interest was expressed in a number of ways by participants, all suggesting a new, common awareness that Argentina and Brazil must achieve greater integration of their economic and security interests if they are to be successful in managing their affairs at home and in projecting power and influence outside the region.

It is particularly significant, therefore, that the two presidents announced an agreement that effectively places the normalizing of nuclear relations high on the list of matters to be resolved. They agreed to work out a system of reciprocal inspections of each other's nuclear facilities, to be complemented by international inspections of *all* their plants to verify the findings. They also agreed to ban nuclear explosions and to bring into force a regional nuclear non-proliferation treaty, the Treaty of Tlatelolco, that had been a dead letter for more than 20 years because of their refusal to adhere to it.

This book presents the conference that provided the first open discussion of the new approaches reflected in this historic nuclear agreement. The conference discussions are summarized in Chapter 1. Formal presentations and panelists' responses are presented in subsequent chapters. These exchanges are significant not only for identifying the problems and the mindset that had to be overcome to reach the agreement but also for identifying the pitfalls to be avoided in its implementation. A number of the papers were originally presented in Spanish or Portuguese and have been translated into English with minimal editing on an *ad verbatim* basis.

At the conference, 'Latin American Nuclear Cooperation: Prospects and Challenges,' nuclear leaders from Argentina, Brazil and Uruguay met for three days (11–13 October 1989) with an independent group of experts from the United States to discuss a number of difficult issues. These were issues that had fuelled nuclear rivalry in the region and had bedevilled non-proliferation diplomacy with the US and other outside powers over the past two decades. Indeed, at times the conference seemed to serve as a catharsis for sharply conflicting views, making possible constructive exchanges of ideas and information after the air had cleared.

Among the issues addressed were the following:

- the appropriate role of atomic energy in the region and, in particular, the value of indigenous efforts by Argentina and Brazil to develop militarily significant technologies for producing enriched uranium and separated plutonium;

- the refusal of the two nations to ratify the Nuclear Non-Proliferation Treaty (NPT) or to submit to inspections by the International Atomic Energy Agency (IAEA) on the full scope of their nuclear activities;

- the denial of significant nuclear technology and materials to Brazil and Argentina by the United States and other advanced industrial nations in the absence of NPT or IAEA full-scope safeguards commitments;

- the question of whether significant nuclear transfers would be forthcoming even if the two nations were to subscribe to the NPT or to comparable measures to demonstrate they were not pursuing nuclear weapons;

- the question of whether the denial of nuclear technology to the region was symptomatic of a larger problem in north-south relations: a more general denial of advanced 'dual use' technologies (those with potential military applications) the dissemination of which could undermine the dominance of the US and other major powers;

- the choice between mutual trust and strict verification in any bilateral arrangements worked out by Argentina and Brazil to assure one another that their nuclear programmes were entirely peaceful;

- the question of whether and how to permit verification by outside parties, especially potential nuclear suppliers, of the peaceful nature of their nuclear activities;

- the question of whether bilateral or international safeguards arrangements could be meaningful unless effective domestic nuclear safeguards and oversight procedures were first established by the civilian governments over their respective nuclear programmes;

- the extent to which development of nuclear-powered submarines in the region contributed to the risk of nuclear weapons proliferation;

- the need for 'peaceful nuclear explosions' and the suitability of the Tlatelolco Treaty – which, by some interpretations, permits such tests – as an alternative to the NPT;

- the extent to which linkages should be made between 'horizontal' and 'vertical' non-proliferation efforts – that is, the extent to which commitments by Argentina and Brazil not to acquire nuclear weapons should be predicated on commitments by the superpowers to do away with their nuclear arsenals.

Differences over these issues had proven intractable in the past. But the conference was intended to build upon progress recently made. Argentina and Brazil had begun to resolve differences between themselves and with the United States and, in particular, to make their nuclear programmes more 'transparent' to one another.

The question yet to be answered was whether they could find a formula, not tied to the NPT but still compatible with the global non-proliferation regime, that would provide concrete assurances that their nuclear activities were peaceful.

The Institute received encouragement and assistance from two prominent research organizations, the Argentine Council on International Relations (CARI) and the Brazilian Institute for the Study of International Relations (IPRI) – each with close ties to the Foreign Ministry in its respective country. CARI and IPRI formally endorsed the conference, and their representatives helped in the development of the programme, the selection of experts, and themselves participated in the meeting.

The stage for the conference had been set by remarkable progress made toward normalizing nuclear relations over the previous four years by President Raul Alfonsín of Argentina and President José Sarney of Brazil – the first civilian leaders after years of military rule in both countries. Following an agreement in 1985 to begin exploring approaches to closer nuclear cooperation, they took turns making 'friendly visits' to the other's most sensitive nuclear facilities – the uranium enrichment and plutonium separation plants.

The visits were mostly symbolic – no secret details on plant design or materials inventories were exchanged. They were significant nonetheless because these plants represented the cutting edge of the Argentine-Brazilian nuclear competition, and the visits were the result of careful preparations by a joint working group under the auspices of the foreign ministries. By 1988, the working group had been elevated to a permanent committee, and a committee of Brazilian and Argentine nuclear industrialists was also formed to 'promote the integration' of their industries.

With the objective of building upon this momentum, the Nuclear Control Institute organized the Montevideo conference to explore specific ways in which the growing nuclear cooperation between Brazil and Argentina could evolve into formal arrangements of lasting value to regional and global nuclear non-proliferation efforts. The conference was a follow-on to a smaller meeting, held in May of 1988 in Buenos Aires at the invitation of CARI, at which the findings of the Institute's International Task Force on Prevention of Nuclear Terrorism were presented and discussed.[1] It was agreed at that session that a larger meeting should be planned to explore a wider range of Latin American nuclear issues.

As the papers and discussions in this book make clear, Argentina and Brazil seem prepared to normalize their nuclear relations, but they strongly resist outside advice on how to go about it. Especially sensitive is the question of making their peaceful nuclear intentions subject to verification by outside parties. A number of Argentine and Brazilian participants refused to acknowledge that *any* form of inspection or verification was needed; mutual trust, they argued, was sufficient. The long history of US intervention in Latin America – and, in particular, the 'policy of denial' that has governed US nuclear trade and non-proliferation efforts in the region – has left scars that were all too evident at the conference.

A Brazilian participant,[2] responding to proposals made by US safeguards and verification experts for strict measures to confirm the absence of nuclear weapons activities, likened such suggestions to advice that a friend might give to a happily married man who trusts his wife: line up a divorce lawyer and a private detective in advance just in case your wife proves to be unfaithful. An American participant responded with an analogy of his own: Argentina and Brazil exempting themselves from full-scope safeguards was like two passengers about to board a crowded plane who refuse to put their hand luggage through an X-ray machine because they worked out a private arrangement to trust one another.

A few Brazilian participants who were not in the government expressed scepticism about the peaceful nature of the Brazilian nuclear programme and about the motives of Brazil and Argentina in normalizing nuclear relations. They argued that in the absence of an effective *domestic* safeguards system with full accounting of nuclear activities and materials to the Brazilian Congress, there was no way to know whether the nuclear programme run by Brazil's Navy included weapons activities. One Brazilian asserted that nuclear cooperation between Argentina and Brazil should be seen as an effort to build trust between their military commands – a form of deterrence – and to build a common front against outside efforts to prevent them from developing their autonomous nuclear capabilities.

These assertions were vigorously disputed by representatives of the Brazilian government and atomic energy programme, who denied any weapons activities and advocated mutual trust as the basis of Argentine-Brazilian nuclear cooperation. They strongly resisted suggestions by American participants for the application of intrusive measures based on existing models – international safeguards as applied within the European Community and bilateral verification as applied by the US and the USSR under the INF (Intermediate Nuclear Forces) elimination agreement. Indeed, one Brazilian branded as 'neo-colonialist' an American suggestion that a bilateral arrangement should have three elements – on-site inspection, surveillance by aircraft or satellite, and third-party verification.

There were also strongly contending views on whether production of weapons-grade materials – highly enriched uranium and separated plutonium – was prudent or necessary, even under safeguards arrangements. One American asserted that safeguards were not too intrusive, as suggested by some Brazilian and Argentine speakers, but too 'slow,' making undetected diversions of bomb materials a real possibility. For that reason, he said, it was dangerous to have unhampered use of these materials.

Some Argentine and Brazilian participants regarded such suggestions as evidence that the United States and other outside powers sought to maintain a monopoly over the technology for producing these materials and over the materials themselves. One American asserted that such materials have proven to be uneconomical and unnecessary in power and research programmes elsewhere and, since Brazil and Argentina are not known to have produced these materials yet in significant quantities, Latin America still could be spared the burden of protecting and accounting for them if production were suspended.

The production of enriched uranium to fuel nuclear submarines being developed by Brazil was discussed in the larger context of whether nuclear submarines contribute to nuclear weapons proliferation. It was noted that only the United States and the United Kingdom are known to use highly enriched, bomb-grade uranium as submarine fuel; France and the Soviet Union are understood to use lower-enriched uranium unsuitable for weapons in at least some of their submarines. Although Brazilian participants did not discuss the level of enrichment of fuel being developed for submarines, there have been official Brazilian statements indicating it would not be highly enriched.

One problem, however, is that low-enriched uranium can be further enriched to weapons grade. Under the NPT, use of submarine and other naval-propulsion fuel is a 'non-proscribed military activity' permitting removal of the fuel from safeguards. Such an arrangement is likely to be included in any non-NPT safeguards arrangement worked out between Argentina and Brazil, as well. Nonetheless, most conference participants regarded a conventionally armed nuclear attack submarine, compared with a nuclear weapon, as the lesser of two evils – or, as an American expert put it, 'Better a sub under the sea than a bomb in the basement.' In any event, conference discussions indicated that an operational Brazilian nuclear submarine was at least 10 to 20 years away and that the Argentine programme has been suspended for cost reasons.

There was also general agreement among participants that another matter of potential proliferation concern – so-called 'peaceful nuclear explosions' – made no economic sense and could be banned from the region. There was general support from Argentine and Brazilian participants to a suggestion by an American conferee that the two nations agree to a *bilateral* comprehensive test ban and challenge the superpowers to join.

The issue of nuclear explosions was also seen as significant to the question of Brazilian and Argentine ratification of the Treaty of Tlatelolco in lieu of the NPT. The Tlatelolco Treaty, by some interpretations, permits the explosion of indigenously developed nuclear devices, which from a technical standpoint are indistinguishable from nuclear weapons. For this reason, outside industrial nations have not accepted the treaty as the basis for technology transfers; nor has the IAEA approved safeguards agreements based on the treaty. There was discussion of the possibility of modifying the Tlatelolco Treaty to bar test explosions as part of the process of Argentina and Brazil ratifying it.

Although there was strong support among Brazilian and Argentine participants for this regional non-proliferation accord, several of them expressed strong opposition to the Nuclear Non-Proliferation Treaty and to the global NPT safeguards regime. They described the NPT as discriminatory for locking in the monopoly of the nuclear-weapon states – 'disarming the disarmed', as one Argentine diplomat expressed it. They also objected to other asymmetries: full-scope safeguards required of non-weapon states while the nuclear-weapon states are free to operate unsafeguarded plants; deployment of US and Soviet nuclear weapons in non-weapon states in Europe – 'geographical' or 'indirect' proliferation, as it was variously described.

Another sore point was the denial of nuclear technology by advanced industrial states to certain non-weapon states party to the Treaty even though the NPT provides for non-discriminatory transfers among Treaty parties. Some Argentine and Brazilian participants were doubtful that their nations would qualify for certain transfers even if they were to ratify the NPT.

American participants noted that technology transfers applicable to production of weapons-usable plutonium and enriched uranium would pose a problem anywhere. Transfers related to power and research programmes utilizing non-weapons-usable materials should not be a problem, they said. They were also generally sympathetic with criticisms that the United States and the Soviet Union could do more to meet their obligations under the NPT to end the nuclear arms race – in particular, by halting nuclear testing and any further production of nuclear-weapon materials.

In any event, the discussions concentrated on the feasibility of developing an effective regional non-proliferation regime *outside* the NPT context. And, in this regard, the heat produced in the discussions generated some light. Proposals put on the table relating to safeguards and verification, to nuclear explosions and the Tlatelolco Treaty, are reflected in the agreement signed at Foz de Iguacu by Presidents Carlos Menem of Argentina and Fernando Collor of Brazil – the elected civilian successors to Presidents Alfonsín and Sarney respectively.

It is easy to be cynical and dismiss the progress made as merely kow-towing to the United States, Germany and other nuclear suppliers that now demand full-scope safeguards as a condition of supply. Some might even see the potential for an Argentine-Brazilian condominium to secretly produce and confront the world with nuclear weapons. On the other hand, it is important to consider the potential for enlightened self-interest that could transform Latin America into a region permanently free of nuclear weapons and could serve as a model for other regions, like South Asia and the Middle East, now troubled by dangerous nuclear rivalries. It is our hope, and our purpose in producing this book, that enlightened self-interest will prevail.

PAUL LEVENTHAL and SHARON TANZER

1 Rapporteur's Summary

Sharon Tanzer

PANEL ONE. WHAT ARE THE GOALS OF ARGENTINE-
BRAZILIAN NUCLEAR COOPERATION AND WHAT
SHOULD BE THE POLITICAL APPROACHES
TO ATTAINING THEM?

Overview

The origins and objectives of nuclear cooperation in Latin America were reviewed by the panelists. Peace, security and development are its goals, said Ambassador Julio Carasales, emphasizing that nuclear cooperation is part of a broader programme of economic integration that could lead to the establishment of a Latin American common market. In contrast, Dr Oliveiros Ferreira viewed nuclear cooperation as an effort by Argentina and Brazil principally to overcome international pressure to sign the Nuclear Non-Proliferation Treaty (NPT) and thereby permit them to proceed with indigenous nuclear programmes, free of international constraints.

The panelists also considered the merits of 'cooperation' versus 'control' as the foundation for an enduring relationship. Dr John Redick proposed the European Atomic Energy Community (EURATOM) as a model for regional nuclear development and advocated safeguards as the best way to ensure that nuclear cooperation will survive the political vicissitudes of democratic government. Ambassador Hector Gros Espiell said nuclear cooperation between Argentina and Brazil was essential to the peaceful use of nuclear energy in Latin America and said an acceptable system of international controls must be found.

Presentations

Ambassador Julio Carasales
Nuclear cooperation between Argentina and Brazil is part of a larger effort to bring about economic integration in the region. This promotion of economic integration is 'the most important political-economic phenomenon' to take place in South America in recent times. The formation of a Latin

9

American common market, should it occur, would be 'the most important event in this region in this century', for it would eliminate any prospect of armed conflict in the region. Nuclear cooperation between Argentina and Brazil must be considered within this framework. Nuclear cooperation is unique because of its special implications for national defence and security. It requires an atmosphere of trust and a broader framework of economic integration. Therefore it is significant that nuclear cooperation has been the most successful aspect of the *rapprochement*.

Argentina and Brazil are committed to developing peaceful nuclear energy programmes independent of the international controls that each regards as discriminatory and as a constraint on its freedom of development. Each nation plans to develop fully the nuclear fuel cycle.[1] Until 1980 Argentina's nuclear programme was more advanced than Brazil's; today the two programmes are in close proximity.

The two nations' long history as rivals for Latin American supremacy have made cooperation difficult, especially in an area as sensitive as nuclear energy. After false starts in the 1950s and '60s, a new attempt at *rapprochement* was made possible in 1979 by settlement of a dispute over hydropower generation along the Paraná River.

In May 1980 the two governments agreed on specific areas for nuclear cooperation in a declaration that emphasized the peaceful nature of their nuclear programmes and their opposition to the development of nuclear weapons. At that time Argentina's National Atomic Energy Commission and Brazil's National Commission for Nuclear Energy also signed an agreement to co-operate in basic research. However, political and economic difficulties, and the complicated transition from military regimes to democratic governments, hindered implementation of these agreements. It was not until democratically elected presidents came to power in 1983 in Argentina and in 1985 in Brazil that the process took on a new momentum.

The current initiative dates from November 1985, when Presidents Raul Alfonsín and José Sarney signed a Joint Declaration on Nuclear Policy at Foz de Iguazú, intended to promote the peace, security and economic development of the region. This declaration endorsed the peaceful use of nuclear energy, promised close cooperation between the two countries' nuclear programmes, and anticipated participation by other Latin American countries in the years to come. As a result of their collaboration, the declaration explained, the two governments will be better able to confront 'the increasing difficulties found in the international supply of [nuclear] equipment and materials'.

To implement the declaration, Argentina and Brazil established a joint working group under the auspices of their foreign ministries. In April 1988 the working group became a Permanent Committee (Declaration of Iperó), to meet every four months. Subgroups were organized in three policy areas: scientific/technical cooperation, foreign policy coordination and legal/technical requirements of cooperation (safeguards). In addition, a committee of Argentine and Brazilian nuclear industrialists, Committee of Argentine and Brazilian Businessmen in the Nuclear Area (CEABAN), was organized to 'promote the integration' of their respective nuclear industries. Two new protocols signed in 1986 dealt with providing assistance in a nuclear emergency and further defined specific areas for mutual cooperation.

Visits by the presidents of Argentina and Brazil in 1986, 1987 and 1988 to each nation's most sensitive nuclear facilities – Brazil's Aramar uranium enrichment plant and Argentina's Pilcaniyeu enrichment plant and Ezeiza plutonium processing facility – underscored the seriousness of the *rapprochement*. High-level visits to sensitive nuclear facilities of a foreign nation are 'truly significant'.

Cooperation by rival nations in an area as sensitive as nuclear energy is a unique experiment. The present Argentine-Brazilian effort can be compared with the Franco/German Coal and Steel Community founded after World War II and the subsequent organization of the European Economic Community. Cooperation has proven more durable than sceptics anticipated; it has survived the election of a new president in Argentina and will survive the election of a new president in Brazil.

International attention has focused on whether cooperation will lead to the creation of an inspection and verification system that would guarantee that neither country is working to produce nuclear weapons. Argentina and Brazil have chosen instead to rely upon the 'mutual trust' that is an outgrowth of their cooperation. None of the documents signed by Argentina or Brazil mentions either safeguards or a system of mutual inspection. The international perception that this agreement is to establish a bilateral inspection system is mistaken, especially when considered in the framework of the NPT, Latin America's nuclear Treaty of Tlatelolco or the International Atomic Energy Agency (IAEA).

Argentina and Brazil have an 'inalienable right to develop, without restrictions, their nuclear programmes for peaceful purposes', as provided in the 1988 Declaration of Ipero. Each nation must decide for itself what security arrangements it considers necessary. The cooperation that Argentina and Brazil have begun addresses their legitimate security concerns. They alone can decide how these concerns are to be met. Reciprocal security assurances, not assurances to the international community, are the goal. However, this process is 'dynamic, not static' and may evolve over time.

Excessive concern by those outside the region with Argentine-Brazilian security arrangements reflects unwarranted scepticism about the peaceful intentions of both nations' nuclear programmes. These concerns fail to consider the broader economic context of this *rapprochement* and assume instead that its objective is to establish a system of mutual inspection.

International non-proliferation agreements do not provide universal solutions. Each nation's special requirements must be considered, in light of its own political, technical and geographic circumstances. Nations must decide for themselves what arrangements they find satisfactory. For Argentina and Brazil the goals of *rapprochement* remain 'peace, security and development'. The realization of these goals will benefit Argentina and Brazil, other Latin American nations and the entire world.

Dr Oliveiros Ferreira

The goals of Argentine-Brazilian nuclear cooperation are political and diplomatic, not economic. Their objective is to deflect US pressure to sign the NPT and thereby make it possible for both nations to continue to develop indigenous nuclear programmes.

To understand what is taking place, it is necessary first to understand the United States' role in Latin America. Historically the United States has held the balance of power in the 'complicated relationship' between Argentina and Brazil; it has kept both countries from dominating the region. If the two nations are to come to an understanding, they must first 'exorcise the U.S. ghost'.

What is the purpose of their nuclear initiative? It is to defend their common international interests by co-ordinating their political positions. Without US pressure to sign the NPT, Argentina and Brazil would not need to co-ordinate their policies.

Rapprochement, however, is easier said than done. Some 150 years of mutual suspicion are not easily overcome, especially by the military commands if not the foreign ministries. Because future economic cooperation depends on the success of nuclear cooperation, the military holds the key to the future. It is the military that controls the nuclear programmes.

Brazilian liberals have looked to the United States as a role model, but it is Brazilian nationalists who have held political power in recent years. In Argentina's nationalism, Brazilians see a common rallying point for a regional foreign policy that is independent of the United States.

To understand nuclear cooperation between Argentina and Brazil, it must be seen as an effort to build trust between the two nations' military commands. The process is more easily understood as a form of 'deterrence'. Argentina and Brazil will share information about their respective nuclear capabilities. Together they will work out a common front against those who seek to prevent them from developing their autonomous nuclear capabilities.

Why is Washington concerned? It may not want to relinquish the balance of power in the region. Washington may be concerned that Argentina and Brazil will use nuclear weapons against each other and destabilize the US rearguard position. The United States may be alarmed by the development of nuclear submarines that could be used outside the region. For all these reasons it is putting pressure on Brazil to sign the NPT.

Finally, Argentina and Brazil must be allowed to work out their own security arrangements, independent of the IAEA, the NPT and the Treaty of Tlatelolco.

Dr John Redick

EURATOM is an appropriate model for nuclear cooperation in Latin America. Although, as Ambassador Carasales asserts, undue attention has been given to the issue of safeguards, nonetheless safeguards or a mutual inspection regime could help ensure that cooperation survives the 'political shifts and changes' of democratic governments. Argentina and Brazil are already exchanging information and research about monitoring equipment, fuel burn-up and other systems that, in fact, are the basis for a safeguards regime.

The next five years will be a challenging time for the NPT regime, whose future will be decided in 1995. Argentina and Brazil should be prepared to offer creative and constructive suggestions for changes in the non-proliferation regime, not just criticism.

Argentina and Brazil played key roles in the negotiation of the Tlatelolco Treaty but have not ratified it. The two nations should revisit the treaty with an 'open mind for compromise', in consideration of efforts now underway to make the Tlatelolco Treaty more attractive to the two nations in the hopes that they may become full parties. A key obstacle to bringing the treaty into force has been the provision for 'peaceful nuclear explosions'[2] (PNEs) and related safeguards difficulties.

With the PNE problem in mind, and with profound implications outside the region as well, Argentina and Brazil could capitalize on the upcoming 1990 Partial Test Ban Treaty conference by announcing a *bilateral* comprehensive test ban and by challenging the superpowers to join.

Ambassador Hector Gros Espiell

Nuclear cooperation is essential to stability and peace in Latin America. However, nuclear cooperation between Argentina and Brazil must take place within the broad framework of economic cooperation. Uruguay has a special role to play in this process because of its close historical ties to both Brazil and Argentina.

The legal instruments that have defined Argentine-Brazilian nuclear cooperation over the years have clearly stated its peaceful objectives. These joint declarations, although not treaties, have an international legal significance and represent the two countries' obligations to each other. Furthermore, as signatories of (although not parties to) the Treaty of Tlatelolco, Argentina and Brazil are obligated to act in a way that is consistent with the objectives and purposes of the treaty.

The bilateral arrangement between Argentina and Brazil must be made subject to a system of international controls. The controls need not be based on the Tlatelolco Treaty or on any existing treaty. However, 'we must ensure that the process is part of an enormous effort for peace'.

Discussion

The discussion focused on the origin of nuclear cooperation between Argentina and Brazil, new policies on PNEs and prospects for the NPT and Tlatelolco Treaty.

An Argentinian challenged EURATOM's suitability as a model for Latin American nuclear cooperation. The nations that make up EURATOM enjoy comparable levels of nuclear and technological development. In Latin America, nuclear development is very uneven, a situation that makes cooperation more difficult. Further, proposals to modify the NPT are unrealistic. In reality, the treaty cannot be amended and attempts to do so are 'a waste of time'.

Another Argentinian called upon the international community to revise its views about the relationship between Argentina and Brazil in consideration of recent political developments in the two countries. To regard the two nations as rivals may have been appropriate when military regimes governed, but today Argentina and Brazil are firmly rooted democracies with *transparent* political processes. Legislatures now have a role to play in stimulating and controlling the power of the executive.

New ways of thinking are necessary to keep pace with a changing world. Conflicts will not occur as they have in the past. Regional conflicts will be different. The East-West conflict is different. The doctrine of deterrence will change. The new reality requires a new approach. The NPT concepts are outmoded and out of step with the new realities. A new language is needed that will employ such terms as 'confidence', 'interest' and 'transparency'. In the future the relationship between Argentina and Brazil will come to resemble that between Belgium and Holland, in which military confrontation does not make sense.

The real issues today have to do with advanced technology, not with nuclear energy *per se*. The same problems of withholding technology that are seen in the non-proliferation regime are present in space research, biotechnology and chemistry.

A Brazilian pointed out the difficulty today of drawing a line between peaceful and military applications of advanced technologies. Although concern exists elsewhere about controls over dual-use technology, Argentina and Brazil are emphasizing mutual cooperation rather than controls.

The speaker endorsed Ambassador Carasales' remarks on nuclear cooperation evolving within a broader framework of economic integration. Confidence-building is one objective of mutual cooperation; a second is to make the nuclear programmes of Argentina and Brazil complementary in terms of material, equipment and technology and to develop together some items that have been denied them in the past 'for reasons of control'.

An American questioned whether economic cooperation must precede nuclear cooperation. In the case of the United States and the USSR, nuclear cooperation is building confidence between the superpowers.

An Argentinian recalled that the incentive for cooperation had its origins in Brazil and Argentina's nuclear programmes, which had run into difficulty as a result of international nuclear non-proliferation policies. The military governments of the time met to explore how they could overcome these problems. Now advanced technology is the issue, not peaceful use. No one believes Argentina or Brazil wants nuclear weapons; the superpowers have proven nuclear weapons are useless in resolving disputes. A Latin American nation that acquired nuclear weapons would be isolated from its neighbours, which have agreed to keep Latin America free from such nuclear weapons. Furthermore, nuclear weapons would jeopardize the significant investments the two nations have made in nuclear power. The basis of the new relationship between Argentina and Brazil is 'mutual confidence', which is more important than treaties.

Signing the NPT is not sufficient to demonstrate peaceful intentions. For example, no one is turning over sensitive nuclear technology to Libya or Iran, although both are NPT signators. Rather, confidence in a nation is what is important. Argentina and Brazil are building mutual confidence daily through technical and industrial exchanges, student exchanges and so forth.

The Argentine speaker was asked to consider the comments that nuclear weapons are useless and their development is contrary to the policies of both Argentina and Brazil and to comment on PNEs.

The subject of PNEs is connected with the Treaty of Tlatelolco. Because Tlatelolco formally permits PNEs, the industrialized nations have not considered it to be a sufficient guarantor of peaceful use and therefore would not accept it as the basis for technology transfers. Furthermore, the IAEA Board of Governors would not approve safeguards agreements based on the Tlatelolco Treaty because it explicitly permits PNEs.

PNEs have not lived up to their original promise. 'Right now, Argentina and Brazil do not believe that peaceful nuclear explosions have any economic usefulness, but it is hard to eliminate this clause from the Treaty.'

A Brazilian participant said that Redick's proposal that a bilateral test ban treaty be considered now, given the evolving relationship between the two countries, sounds interesting and requires further study.

Another Argentinian emphasized that it is the nuclear weapon states that have been conducting nuclear tests; their attitude toward a nuclear test ban is more important than Argentina's or Brazil's. If there were an agreement to ban all nuclear tests, Argentina and Brazil would participate.

An American participant recalled US-Soviet test ban negotiations and the role of PNEs. A US government-industry programme to test the commercial viability of PNEs led to the conclusion that they were 'economically unsound'. Eventually the Soviets agreed to a ban on PNEs. An attractive feature of the CTB proposal was that it was 'basically non-discriminatory. There were no nuclear weapon states, no non-nuclear weapon states. There were simply the states of the world'.

A Brazilian noted that confidence-building measures are in place between Argentina and Brazil, and between the United States and the USSR. What is needed now are confidence-building measures between the nuclear weapon states and the non-nuclear weapon states. The conference to amend the Partial Test Ban Treaty (PTBT) will provide an opportunity to gauge the attitude of the nuclear weapon states on this important issue.

PANEL TWO. WHAT ARE THE INDUSTRIAL AND
ECONOMIC BENEFITS OF LATIN AMERICAN
NUCLEAR COOPERATION?

Overview

This panel reviewed the history of nuclear energy programmes in Brazil
and Argentina, their development of indigenous nuclear fuel cycle facilities, the measures underway to integrate their nuclear industries and their
prospective markets. The panelists also discussed the relative advantages
of developing an indigenous nuclear industry as against an industry based
on the import of state-of-the-art nuclear equipment and technology. The
latter policy choice would require at least the acceptance of international
safeguards on all imported nuclear technology and facilities and on the
nuclear material produced through their use.[3]

Presentations

Fernando Henning
Henning reviewed the history of Brazil's nuclear energy programmes.
Brazil has followed a two-track policy of nuclear development – importing
nuclear reactors and nuclear technology and at the same time actively
developing indigenous nuclear fuel cycle facilities. In the 1950s and
1960s Brazil imported two research reactors from the United States; a
third US-Brazilian research reactor began operation in 1965. In 1968 Brazil
ordered its first nuclear power reactor from Westinghouse.

In 1975 Brazil announced a comprehensive agreement with the Federal
Republic of Germany (West Germany) to provide Brazil with nuclear
power reactors and nuclear fuel cycle technology. The agreement called
for the construction of eight power reactors, with progressively increasing
participation by Brazilian companies. West Germany also agreed to supply
jet-nozzle enrichment and fuel-fabrication technology. In addition, the two
countries agreed to co-operate to confirm the existence of uranium reserves
in Brazil.

Brazil's 'autonomous' or 'parallel' programme has focused on developing the fuel cycle technology that Brazil could not import, in particular,
ultracentrifuge enrichment technology. Brazil decided to proceed with an
autonomous nuclear programme because the jet nozzle enrichment technology supplied by West Germany had never been tested on a commercial
scale. Another factor was the US decision to cut off further supply of

nuclear fuel to Brazil's Angra 1 power reactor in 1978 when Brazil refused to accept the full-scope safeguards required by the US Nuclear Non-Proliferation Act. Brazil has now successfully demonstrated operation of an indigenous pilot centrifuge enrichment plant.

A major reorganization of Brazil's nuclear programme in 1988 gave responsibility for all fuel cycle projects, including those derived from the agreement with West Germany, to the National Commission on Nuclear Energy (CNEN). ELETROBRAS took over all nuclear power reactor activities.

Brazil's ambitious nuclear programme, conceived in a period of economic expansion, was curtailed by the economic crisis of 1983. New reactor construction was cancelled, and ongoing projects were delayed.

Nuclear cooperation between Argentina and Brazil is intended to take advantage of the complementary strengths of the two nations' nuclear industries, which are both expected to participate in the construction of Argentina's Atucha II and Brazil's Angra 2 nuclear power reactors. The 1985 Foz de Iguazú Declaration called for such industrial cooperation; a working group of nuclear industrialists was organized to implement it. The group considered 'mechanisms' to verify the peaceful use of nuclear energy but abandoned this effort in favour of 'mutual trust' arrangements.

The programme of nuclear cooperation allows for participation by other Latin American nations, but the different levels of nuclear development in the region preclude this step for now.

Latin American economic integration will ultimately depend on economic growth in the region. This growth cannot occur without a resolution of the foreign debt problem, without changes in the industrialized nations' discriminatory trade practices and without the adoption of deep structural reforms in the Latin American economies. However, the decision to establish a Latin American market is one for the governments of the region to make, which can then encourage participation by the private sector.

Nuclear energy is, in some ways, unique. The industrialized nations have withheld access to advanced technologies not only for non-proliferation reasons, but also because of their commercial and political interests. Today, nuclear energy is of vital importance for industrial development. Environmental restrictions will increasingly curtail the use of hydropower and fossil fuels. Regional cooperation on the development of nuclear energy makes economic sense because of the high cost of developing indigenous nuclear technology. Nonetheless, in the years to come, cooperation with the industrialized countries that dominate nuclear technology will remain indispensable for Latin America.

José Bernal Castro

Bernal Castro reviewed the history of Argentina's nuclear programme, which has been under the direction of Argentina's National Atomic Energy Commission (CNEA) since it began. The CNEA reports directly to Argentina's president.

Argentina's nuclear programme began in the 1950s with the training of personnel. Mining and mineral production were begun in the 1960s. In 1974 Argentina's first nuclear power reactor (Atucha I) began operation; a decade later a second reactor (Embalse) started up. The growing strength and self-sufficiency of Argentina's nuclear industry is demonstrated by the fact that although Atucha I was an imported 'turnkey' operation from West Germany, Embalse was imported from Canada but built with substantial contributions by domestic engineering and construction firms. The domestic component will be even greater for Atucha II, an import from West Germany, whose construction has been delayed by Argentina's economic crisis.

In the past decade Argentina has demonstrated its mastery of the nuclear fuel cycle, as evidenced by the construction and operation of an enrichment plant at Pilcaniyeu. However, the economic crisis has forced a severe curtailment of nuclear research projects.

Nuclear energy development in the region is very uneven – only Argentina, Brazil and Mexico have nuclear power reactors. This situation is an incentive for integration, however, and not a drawback. Argentina has agreements for nuclear cooperation with eleven Latin American nations. Argentina's principal contribution to regional nuclear development has been in nuclear education and nuclear research. Argentina has exported a research reactor to Peru, a noteworthy event because of Peru's political decision to purchase a research reactor from a Latin American country. The two nations have worked closely on this enterprise.

Argentina and Brazil have had to overcome a long history as regional rivals in order to build a co-operative relationship. The change that began in 1980 is the result of two factors: a new appreciation of the benefits of regional cooperation; and a desire to refute the superpowers' contention that Argentina and Brazil were engaged in an arms race, demonstrating instead a programme of nuclear cooperation. By doing so Argentina and Brazil have shown that it is possible to co-operate in the development of nuclear technology.

Economic integration is critical to economic growth in Latin America. Economic integration is a reality for the industrial powers, and it is the only sensible course for Latin America. The nuclear industries of Argentina and Brazil must work together to develop their industrial capacity, with the support and encouragement of both governments. Once Latin American nuclear integration has become a reality, the industrialized nations will have to recognize that nations in this region also have the right to establish their own criteria for growth.

Samuel Edlow

Edlow questioned the economic and industrial value of nuclear cooperation between Argentina and Brazil, as presently structured, while acknowledging its political and philosophical value. Because cooperation is now confined to the development of *indigenous* nuclear technology and facilities, it is too limited to have significant economic and industrial value. An evaluation of their programme shows that Argentina and Brazil are using scarce economic resources to 're-invent the wheel'. They would be better advised to import state-of-the-art nuclear technology and equipment, improve it and adapt it to local needs. However, until Argentina and Brazil are prepared to place their entire nuclear programmes under international safeguards, they will not have access to the global nuclear market. The cost of Argentine-Brazilian nuclear cooperation as presently structured will be enormous and the economic benefits insignificant. The two governments will have to accept this reality before they can be in a position to take advantage of the full potential of nuclear cooperation.

Dr Walter Cibils

Cibils stressed the importance of viewing nuclear relations between Argentina and Brazil in the context not only of Latin American nuclear development but also of overall regional integration. It is only through economic integration that South America can improve its position in the international community, as suggested by the examples of the European Economic Community and the US-Canada free trade agreement.

The countries of the region must determine what is the most appropriate path of development for them. That path may not be the same one followed by the industrialized countries. However, nuclear technology, a 'cutting edge' technology, could be a valuable tool for economic integration, particularly for those countries not blessed with abundant natural resources.

Countries such as Uruguay can learn a great deal from the experience of Argentina and Brazil, both from the great effort those two countries have made to share their technology and from the mistakes they have made in the course of developing it. However, for Uruguay and many other Latin American countries, human resources will be the decisive factor in their economic development.

Discussion

A Brazilian participant questioned the accuracy of Henning's account of Brazil's nuclear development. Brazil's nuclear programme failed because it was misconceived and did not respond to Brazil's problems – not because of the 1983 economic crisis. Brazil did not need nuclear power in 1975 when it signed the agreement with West Germany. The Itapu hydroelectric plant then under construction was a top priority of the Brazilian government and has since proven to be a marvellous success. The Brazilian government itself abandoned the original nuclear programme with West Germany in 1980.

The autonomous programme is essentially a programme to develop Brazil's capabilities in the field of nuclear technology. It has to do with national independence and national autonomy rather than the generation of electricity. Therefore serious economic benefits from nuclear cooperation cannot be expected.

There are very major obstacles to economic cooperation in Latin America, particularly economic cooperation between Argentina and Brazil. Where such cooperation could be beneficial, it has not taken place. Brazil has surplus electrical capacity in its southern region that could be used to alleviate Argentina's power shortage. However, the power grids of the two nations have never been joined. Mistrust between the military staffs of the two nations has prevented this cooperation from taking place.

Not everyone agrees that Brazil needs nuclear energy; it is a matter of serious dispute within Brazil, as it is in the United States. The prospects and challenges of cooperation have to do with building confidence, which exists to a much lesser extent than has been indicated here this morning.

Argentine participants challenged Edlow's assertion that nuclear technology would be made available to Argentina and Brazil if they signed the NPT. Since the NPT came into force, the Zangger Committee, the London Suppliers Club and the US Nuclear Non-Proliferation Act have placed additional restrictions on technology transfers. 'Key technologies like reprocessing and enrichment will not be forthcoming.'

A US participant emphasized the need to distinguish between sensitive nuclear technologies for producing weapons-usable materials and nuclear technologies for generating electricity without utilizing such dangerous materials. It is access to the former that is restricted.

The basic purpose of the meeting, he suggested, is to explore whether a bilateral system can be developed – with safeguards and verification arrangements that are acceptable to Argentina and Brazil as well as to third parties. Putting such a system into place could provide the basis for hitherto restricted transfers of technology going forward. In this way it would be possible to avoid disputes linked to the Nuclear Non-Proliferation Treaty (NPT) and its safeguards regime, which are not the subject of this meeting.

An Argentine participant replied that technology transfer issues and the economics of developing an independent technology have implications that extend beyond the nuclear industry. It is really a question of the industrialized north wanting to maintain its markets in the dependent south.

PANEL THREE. WHAT ARE THE AVAILABLE MODELS FOR DEVELOPMENT OF A BILATERAL NUCLEAR CONFIDENCE-BUILDING REGIME?

Overview

The panelists' presentations and the subsequent discussion of a 'confidence-building' regime made clear that there was a wide range of viewpoints on how such a regime should be structured, with consensus reached only on the importance of national safeguards to achieving effective bilateral or multilateral arrangements. Most participants from Argentina and Brazil voiced support for the existing informal confidence-building arrangements between the two countries. A number of US participants, on the other hand, advocated development of a more formal safeguards/verification arrangement to increase 'transparency' of each nuclear programme to the other and to build mutual confidence in the peaceful intentions of both nations. Such an arrangement, some argued, would more likely survive any future political strains and would permit Argentina and Brazil greater access to international trade in nuclear goods and technology.

Presentations

Dr William Higinbotham
Higinbotham suggested that a bilateral arrangement outside the NPT framework could be modelled upon existing international safeguards agreements. He discussed the need for a national safeguards system, reviewed international models for safeguards (IAEA and EURATOM) and proposed less comprehensive bilateral measures. Whatever the safeguards system, it must be freely adopted if it is to be effective; it cannot be imposed. Each party

must conclude that a safeguards agreement will advance its own security and welfare.

Agreements on international safeguards rely upon national safeguards systems that establish and enforce regulations for the control, accounting and physical protection of nuclear material. A national safeguards system is necessary to account for nuclear material and protect it from misuse. An international safeguards system establishes accounting procedures for nuclear material that permit auditing and confirmation of data supplied by the national system.

The United States established civilian control over atomic energy in 1946 when the US Congress enacted the Atomic Energy Act. The experience of the next three decades made clear the need to separate regulatory and promotional activities, to divorce safety and safeguards responsibilities from those for production, and to provide for public review of policies and programmes.

All South American nations with significant nuclear programmes have national safeguards systems. Experts from these countries participate in international seminars on nuclear material control and accounting; earlier this year they contributed to the revision of the IAEA guidelines on physical protection. Their common goal is an efficient and effective national safeguards system.

The objective of the IAEA is to promote nuclear energy and to ensure that the material and technology utilized are not applied to military purposes. In its promotional capacity the IAEA provides technical assistance and training to developing nations. In its safeguards capacity the IAEA defines and applies safeguards.

Most nations are members of the IAEA. A number of government officials at this conference have participated in the IAEA's technical and advisory group meetings and are no doubt familiar with the implementation of IAEA safeguards agreements. A representative of Brazil is a member of the IAEA's safeguards advisory group.

There are two categories of IAEA safeguards agreements: those for states that submit all their nuclear material for IAEA inspection (so-called INFCIRC/153 safeguards); and those for states that submit only some material to IAEA safeguards (INFCIRC/66). The latter are necessarily more complicated. For example, Argentina's heavy water production plant is under IAEA safeguards; if all of Argentina's nuclear facilities were safeguarded, it would not be necessary to safeguard the heavy water production facility. (Heavy water in Argentina must be safeguarded to verify that it is not used in an unsafeguarded reactor.)

A second international safeguards model is offered by EURATOM.

EURATOM was formed in the early 1950s to promote nuclear energy, ensure equal access to resources and technology by member states and provide mutual assurance of peaceful use. Initial plans called for EURATOM to undertake nuclear research and development, manage nuclear power plants and associated facilities, and contract for and own all special nuclear material. The withdrawal of France from EURATOM's joint research and development effort slowed its progress but did not affect the implementation of the safeguards.

EURATOM safeguards inspectors have unlimited access to EURATOM nuclear facilities and personnel. Policies regarding nuclear material accounting are made by the Commission and enforced by member states, which are also responsible for physical protection measures. IAEA safeguards apply along with EURATOM safeguards, although IAEA inspectors do not have the same unlimited access to EURATOM facilities and personnel.

A third option is a more limited programme of mutual assurance between Argentina and Brazil that could include the following steps:

- announcement that the programmes are for peaceful use;

- exchange of visits by officials and experts;

- bilateral inspections;

- multinational operation of sensitive facilities.

Dr Milton M. Hoenig

Hoenig described recent arms control agreements that could serve as potential models for a bilateral nuclear confidence-building arrangement between Argentina and Brazil. Such a bilateral regime could ensure greater 'transparency' in the two nation's nuclear programmes and therefore provide a high degree of confidence in their commitment to peaceful nuclear energy. A bilateral agreement outside the NPT framework could incorporate more stringent measures than are acceptable in an international regime.

Contrary to the expectations of a decade ago, the trend in arms control today is toward greater intrusiveness. A bilateral verification agreement between Argentina and Brazil could have three possible elements: on-site inspection; aircraft and satellite surveillance; and third-party certification.

For effective on-site inspection arrangements, access to facilities is the important issue. Agreements may specify unlimited access, provide for challenge inspections, or permit inspections of a suspect site. The recent INF agreement between the United States and the USSR includes provisions for highly intrusive inspections.

A less intrusive arrangement is verification by aerial and satellite surveillance. The two parties could ask the superpowers to share intelligence from their satellites, they could buy commercial satellite services or they could operate dedicated surveillance systems of their own. Satellites could be used to look for such clandestine activities as enrichment plants, reprocessing plants, production reactors or test sites. They could also identify new construction at established nuclear facilities. Aircraft surveillance would be less expensive and more efficient than satellites, but it is more intrusive and could be a source of friction.

In a bilateral system under which the two countries would exchange inspectors, inspectors of a third country could be invited to 'look over their shoulder' to guarantee the regime's *bona fides*. Sweden, Switzerland, Japan and Australia are nations with the political and technical credentials to provide third-party certification. With such certification it might be possible to open channels to major nuclear suppliers that now withhold exports for lack of full-scope safeguards in Argentina and Brazil. Other possible third-party certifiers are international or regional nuclear organizations, such as the IAEA or OPANAL, or even a private company with a reputation for independence and intergrity.

Argentina and Brazil could take the first steps toward a bilateral regime by establishing a special bilateral commission to look into procedures for safeguards and verification. The commission could examine the feasibility and desirability of: reciprocal, short-notice, on-site inspections; an 'open skies' policy using satellites and overflights; and third-party certification. Recent superpower arms control agreements incorporate all these techniques.

Dr Miguel Estrada Oyuela
Estrada Oyuela explained that Argentina and Brazil believe an informal confidence-building regime is more appropriate today than a formal safeguards regime. Safeguards grew out of the post-World War II European climate of suspicion. They create an adversarial relationship because they require an independent party to verify that no nuclear material has been diverted, presumably for the purpose of constructing nuclear weapons. IAEA safeguards incorporate this philosophy, which has become the basis for the global non-proliferation regime.

In Latin America the situation is completely different from that in Europe. The Treaty of Tlatelolco represents a consensus among Latin American nations to keep the region free of nuclear weapons. In Latin

America a 'climate of goodwill' is the basis for cooperation and mutual confidence. Plans by Argentina and Brazil for joint development of a breeder reactor – a 'sensitive' technology – are an expression of the confidence the two nations have in each other.

The current dilemma of non-proliferation policy is the approach of dealing with the proliferation of technology as if it were proliferation of nuclear weapons.

Dr Bernardino Coelho Pontes

Pontes agreed with Higinbotham on the importance of a national safeguards system: any misuse of nuclear materials is likely to affect the citizens of the nation where it occurs. As for the confidence-building measures proposed by Higinbotham – for example, an exchange of official visits to inspect nuclear facilities – these are the very measures that have been instituted. Bilateral inspections, as proposed by Hoenig, should be rejected. They would constitute a step backward that would damage the credibility of the IAEA as well as the Agency's role in conducting inspections that had previously been conducted on a bilateral basis. There is no need to *develop* confidence, it already exists, as the exchange of visits to sensitive facilities confirms.

The arms control models proposed by Hoenig are impractical and contrary to established legal norms. They would create an International Nuclear Central Intelligence Agency. Pontes observed, 'I had no idea we were so dangerous.'

Dr Gerardo Quintana

Quintana described technical training programmes in Argentina that enroll students from all over Latin America and promote mutual confidence through mutual acquaintanceship. In their ten years of operation these programmes have trained 400 students many of whom are now influential scientists and policy-makers throughout Latin America.

Dr Fernando de Souza Barros

Souza Barros agreed with other panelists on the importance of a national safeguards system. Any bilateral safeguards arrangement between Argentina and Brazil would have to rely on the national systems. The Brazilian Physical Society and the Brazilian Society for the Advancement of Science have asked Brazil's Congress to establish a national system with the technical capability to inspect those parts of the Brazilian nuclear programme that are not under international safeguards.

The time to put a bilateral agreement in place is while relations between the two nations are cordial. If problems arise, the strong reliable national systems will help minimize their severity. As for satellite surveillance or aerial reconnaissance, this approach was unlikely to be acceptable in Latin America.

Discussion

The differing attitudes concerning the purpose and function of a confidence-building regime became clear during the discussion. The conference organizers had stated the issue as follows:

> How do two nations that, for political and national policy reasons choose not to subscribe to the international regime of fullscope safeguards, establish between themselves a system that will provide assurance to one another over the long term, and will provide confidence to interested third parties, perhaps suppliers who under present circumstances are not able to provide supply?

An Argentine participant said that safeguards are necessary only when confidence between parties does not exist. This is not the case today for Argentina and Brazil. Mutual confidence is the basis on which cooperation has been launched. Both nations agree on the need for a co-operative relationship, and together they are moving toward a lasting and enduring integration of their nuclear programmes. Exchange of information about each other's programmes has taken place in the context of presidential visits to sensitive facilities and frequent technical exchanges.

A Brazilian participant confirmed those remarks, stating that a bilateral safeguards regime creates 'suspicion', not confidence.

There was widespread agreement during the discussion that a national safeguards system (nuclear materials accounting and physical protection) is an important element of any nation's nuclear programme and a prerequisite for either multilateral or bilateral safeguards regimes. However, there were strongly felt differences as to whether adequate national safeguards systems already existed or whether they should be applied to bilateral arrangements.

Several Brazilian participants insisted on the need for much stronger civilian control of Brazil's nuclear programme and greater accountability to civilian authorities. They pointed out that Brazil has no independent civilian nuclear oversight outside the executive branch. The Brazilian Physical Society has called for the creation of an independent institution to oversee and to regulate Brazilian nuclear activities and for a national inspection system to operate alongside the international safeguards system at certain nuclear facilities and activities. Without these improvements Brazil's national system cannot be an effective basis for a bilateral or multilateral arrangement. Despite the new Brazilian constitution's assurances that all nuclear activities are for peaceful purposes, some participants from Brazil questioned whether this is actually the case and whether all Brazil's nuclear activities have been made public. There is a great need to 'bring democracy to the nuclear question in Brazil', one asserted. 'For a bilateral agreement to be meaningful, there must be civilian control and a much greater transparency with respect to the roles and purposes of nuclear activities in Brazil.'

Brazil's constitution assures civilian control of the nuclear programme and gives Congress the authority to supervise it, other Brazilians responded. They objected to a discussion of national safeguards systems as inappropriate at a conference at which Brazil-Argentina nuclear cooperation, not individual national programmes, was to be discussed. In fact, third parties have no legitimate interest in a system of cooperation that Argentina and Brazil have established to their satisfaction. It is a bit as if a good friend suggested to a happily married couple that they engage a private detective to spy on each other in case they should someday decide to obtain a divorce, one Brazilian observed.

An Argentine participant suggested that the purpose of international safeguards is different from that of national systems for material accounting and physical protection.

A US participant responded, 'A reliable national system is basic. The IAEA would certainly depend on the national system of accounting.'

With respect to an earlier reference to Brazil's 'parallel' programme, an Argentine participant said Argentina understands it to be 'parallel' to Brazil's agreement with the Federal Republic of Germany. It is a programme to develop indigenous nuclear technology. Argentina is aware of the need for secrecy and developed its own enrichment technology that way. International pressure 'would have been unbearable if it had been public'. Such a programme 'could have awakened a degree of suspicion'. It did not do so because of its 'internal coherence. . . . There is no reason to see [in the parallel program] any kind of second intentions'.

A US participant suggested that unwritten, informal assurances are satisfactory if the objective of Argentine-Brazilian cooperation is to promote nuclear trade between themselves. If, however, Argentina and Brazil wish to participate in a broader, multilateral nuclear forum and to have access to a greater flow of technology and trade, then bilateral expressions of confidence will not be sufficient. A bilateral arrangement without intrusive inspection may satisfy the two parties and the region, but it will not satisfy third parties who must assess whether, in the absence of NPT ratification, nuclear supply should be forthcoming.

The purpose of this meeting, another US participant suggested, is to try to find a middle ground, a bilateral arrangement outside the NPT satisfactory to the parties as well as to outside suppliers.

A Brazilian participant said that Hoenig's proposals for bilateral verification arrangements smacked of 'neocolonialism'. A second Brazilian said that charges of 'neo-colonialism' illustrate the point that Brazilian nationalists are seeking in Argentina an ally against the United States. The nuclear weapon states are afraid of the nuclear programmes of Argentina and Brazil because they are concerned that these programmes could have an impact on the security of other countries. The purpose of this seminar is to learn Argentina and Brazil's reaction to their concern. Brazilians who believe in Brazil's 'manifest destiny' do not like to be told what Brazil must do and do not believe that any nation has the right to be the 'police of the hemisphere, or the bishop'.

A US participant made these final points. Of course, it is up to Argentina and Brazil to decide what it takes for each to have confidence in the other. It is impossible to ignore, however, that what these nations do influences the nuclear decisions of other nations. In addition, others see their unwillingness to accept comprehensive safeguards, applied internationally or bilaterally, as evidence of retaining the option of military use of nuclear energy.

PANEL FOUR. DOES THE ACQUISITION OF NUCLEAR SUBMARINES CONTRIBUTE TO THE RISKS OF NUCLEAR PROLIFERATION?

Overview

Nuclear submarines[4] are relevant to a discussion of nuclear proliferation because of the submarines' requirement for enriched uranium and because of the IAEA provision (INFCIRC/153, Article 14) that exempts nuclear submarine fuel from safeguards. The panelists' assessment of whether submarines represent a proliferation risk differed according to the degree of emphasis they placed on a nation's *technical* capability to produce a nuclear weapons material in relation to a nation's *political* commitment to renounce the acquisition of nuclear weapons. Admiral Carlos Castro Madero said it is the political decision that is crucial. Dr Marvin Miller suggested a nation's technical capabilities can influence its political decision. Dr José Goldemberg pointed out the vital need for a national safeguards system to provide domestic accountability.

Presentations

Dr Marvin Miller

Miller noted that US nuclear submarines use highly enriched uranium (97.3 per cent U-235), a nuclear weapons material. By contrast, French submarines use fuel enriched to less than 10 per cent U-235; Soviet submarines may use comparable enrichments. While not directly usable in nuclear weapons, even low-enriched uranium 'provides the rationale for the establishment of uranium enrichment facilities under national control' which could be converted to the production of highly enriched uranium.

The NPT allows a state either to withdraw safeguarded nuclear material from safeguards for a 'non-proscribed military activity,' such as fuel for a nuclear submarine, or to import unsafeguarded material for this purpose from a nuclear-weapon state or a non-NPT state, without triggering safeguards. The exemption from safeguards also applies to the enrichment and fuel fabrication facilities intended for production of nuclear submarine fuel.

To provide assurance that unsafeguarded material is not being diverted to weapons use while it is withdrawn from IAEA safeguards, a state may choose to negotiate a bilateral safeguards agreement with the supplier. Canada had planned to provide continuity of safeguards coverage in this manner in its nuclear submarine programme. These arrangements suffer from a certain lack of credibility, however, because both parties may have an interest in minimizing any violations that may occur.

The proliferation risks associated with nuclear submarines can be minimized. Highly enriched uranium in fabricated reactor cores could be supplied under stringent safeguards. This situation would be preferable to one in which low-enriched uranium is supplied from an unsafeguarded, indigenous enrichment plant.

Nuclear-powered submarines may prove to be 'an attractive surrogate for nuclear weapons. . . . Better a sub under the sea than a bomb in the basement'. There is a widespread consensus among naval strategists that command of the seas in the future lies with the nuclear submarine. This prediction provides a powerful rationale for the acquisition of nuclear submarines both by militarily significant third world states such as Brazil and India and by members of the western alliance, e.g. Canada, Spain, Italy and Japan. Although recognizing there are proliferation risks associated with a nuclear submarine programme, these risks should not be exaggerated. The number of states with nuclear submarines will remain small. 'There is time to develop a considered policy.'

Admiral Carlos Castro Madero

Castro Madero stated that nuclear submarines do not contribute to the risk of nuclear proliferation. The IAEA specifically permits safeguarded nuclear material to be removed from safeguards and to be used in nuclear submarines. As a consequence of Britain's use of nuclear submarines during the Malvinas War, Argentina requested an opinion from the IAEA as to whether the use of safeguarded nuclear material in nuclear submarines violated the IAEA's peaceful use requirement. Although the IAEA's statute directs that nuclear energy not be used for military purposes, the IAEA replied that nuclear naval propulsion systems are not incompatible with a country's commitment to the peaceful use of nuclear energy. In this connection, Castro Madero cited Canada as a leading advocate of non-proliferation that had also actively explored a nuclear submarine programme.

A nuclear programme can give a nation the 'potential capacity' to produce nuclear explosives. However, it is the *political* decision that counts. Without a political decision, there will be no nuclear weapons programme. Furthermore, no nation has chosen this route to nuclear weapons. A submarine reactor is a compact reactor similar to that of other reactor designs, and it does not have to operate with weapons-grade uranium.

Finally, acquisition of a nuclear submarine is unlikely to trigger a regional nuclear arms race. Such a response would be out of proportion to the event. A nation with the technological infrastructure to produce a nuclear weapon is more likely to respond, if at all, by acquiring its own nuclear submarines.

Dr José Goldemberg

Goldemberg reported on the status of Brazil's nuclear submarine programme and stressed the need for civilian control. His account of the rationale for Brazil's acquisition of nuclear submarines was based on an article by Admiral Mario Cesar Flores in a 1988 Brazilian naval journal.[5] In the years to come, according to Flores' article, Brazil must have a military power commensurate with its growing international presence and responsibilities in the South Atlantic. Nuclear submarines, because of their speed and the distances they can travel without refuelling, are ideally suited to this role. Brazil's 'autonomous' nuclear programme was begun in 1979 in order to develop the technology Brazil would need to build nuclear submarines. The programme was kept secret, according to the Admiral, in order to avoid public opposition. Today, Brazil has a centrifuge enrichment plant in which it takes great pride; secrecy is no longer necessary.

Is this programme a cover to develop a nuclear weapons capability? Although weapons production is technically feasible, Admiral Flores wrote, Brazil has made a 'national decision' to use nuclear energy for peaceful purposes only, including a nuclear submarine programme. In the future this project will be subject to the authorization of Brazil's Congress.

Goldemberg said that Brazil is indeed justified in taking pride in this technical achievement. The nuclear submarine programme, however, is a very long-term one. It will be twenty to thirty years before Brazil will have an operational nuclear submarine, according to Admiral Flores.

According to Goldemberg, the project is also very expensive, and it must be considered in the context of Brazil's other needs, social and economic, as well as technical and scientific. It cannot be given absolute priority but must be considered within a framework of national priorities.

Finally, the programme must be placed under very strict internal safeguards. Internal safeguards in this context means a system accountable to civilian authority and about which civilians are kept well informed. A strong internal safeguards system is the prerequisite for a bilateral safeguards system between Brazil and Argentina. Adoption of any future international safeguards system will require an internal safeguards system to be in place.

Discussion

In the discussion that followed, an Argentine participant commented that Argentina's position with respect to the NPT had nothing to do with the fact that the NPT does not prohibit nuclear submarines. After pointing out the difference between the views of Admiral Castro Madero, who concluded there was no proliferation risk as a result of the development of nuclear submarines and the views of Dr Miller, who concluded there was some risk but that it should not be exaggerated, the Argentine participant agreed with Castro Madero: nuclear propulsion poses no risk of nuclear proliferation. To contend otherwise is to assume that all nuclear activity poses some risk of nuclear proliferation.

Although nuclear submarines are acceptable today, as third world nations seek to acquire them, new prohibitions similar to recent attempts to curb the development of ballistic missiles may emerge. In 1987, without any legal justification, high technology countries began another 'London Club' type arrangement[6] to restrict transfers of missile and space technology. There is now a growing link in the literature between missiles and nuclear proliferation. It would be unfortunate to have the same process occur with nuclear submarines. In this respect it is worth noting Miller's observation that international attitudes toward a new class of states – non-nuclear-weapon states with nuclear submarines – will vary according to the identity of the state and its position on the political and ideological spectrum. Miller's proposition, 'Better a sub under the sea than a bomb in the basement', seems accurate but it should be added that this point was not meant as an endorsement of nuclear submarines. Each country must decide for itself the merits of such a programme.

The conference chairman asked for a discussion of the importance of a national safeguards system in the context of a programme to develop a nuclear naval propulsion system, a point emphasized by Goldemberg. The chairman noted that an NPT state must declare the nuclear material it removes from safeguards; in a non-NPT state the material is not declared in the first place. 'What sort of national safeguard system has to be developed to provide assurances that material being developed for the naval propulsion programme is not of weapons grade or is not being utilised for weapons development?'

A Brazilian pointed out that the Brazilian Nuclear Energy Commission's initial interest in compact nuclear power reactors had to do with energy production, not naval propulsion. The Navy decided to lend its support to assure that Brazil would one day have the means to proceed with the development of a nuclear submarine if international restrictions arose in the future as they had in the past. The Navy has allocated no funds for submarine construction; it is assisting the Brazilian Nuclear Energy Commission in developing a prototype reactor.

Another Brazilian emphasized the importance of a national safeguards system if Brazil does proceed with a nuclear submarine programme because of the substantial amounts of plutonium and highly enriched uranium that could be involved. While it would be a strange way to make a bomb, when it is remembered that Brazil has expended almost $7 billion without producing a single kilowatt hour of electricity, the scenario is not unimaginable. The nuclear submarine programme is under the Navy's control; national safeguards are usually under civilian control. Brazil's Congress therefore must take a strong hand in developing a national safeguards system – a system of national accountability and control.

An Argentinian said that if this discussion suggests that nuclear technology vital to a nuclear submarine programme is indistinguishable from the technology involved in nuclear weapons production, it must then be asked whether it is legitimate to attempt to proscribe the technological progress of developing countries. 'We understand that nuclear technology may present a proliferation risk.' However, nuclear proliferation is a political problem; it is not a technical problem. If the world is going to erect technological barriers to nuclear proliferation, it will be restricting a nation's right to develop peaceful nuclear energy.

The chairman posed two questions: is there a possibility of Argentina and Brazil jointly developing a nuclear propulsion system, including shared technology, for the purpose of regional security against third parties? does Argentina have a programme to develop a nuclear propulsion system?

An Argentinian recalled that after Argentina had requested and received an opinion from the IAEA that the development of a nuclear submarine was not incompatible with a programme for the peaceful use of nuclear energy, Argentina's nuclear energy commission ordered a feasibility study on the possibility of developing such a submarine. This study, completed at the end of 1983, concluded that it was possible for Argentina to build a nuclear propulsion system.

Another Argentinian added that the government gave the green light to proceed but allocated no funds. Therefore nothing has been done.

A third Argentinian recommended that technical studies on nuclear submarine propulsion should continue, but noted that in the current economic situation Argentina faces more pressing needs. As for the possibility of a joint submarine project with Brazil, this expert thought it 'would be very difficult'.

PANEL FIVE. WHAT LINKAGES SHOULD THERE BE BETWEEN HORIZONTAL AND VERTICAL NON-PROLIFERATION?

Overview

The NPT represents a bargain between nuclear-weapon states and non-nuclear-weapon states that links efforts to halt the 'vertical' growth of nuclear weapons and their 'horizontal' spread. In return for a promise by the non-nuclear-weapon states not to acquire nuclear weapons and a commitment to open all their nuclear facilities to international inspection, the nuclear-weapon states pledge to pursue good-faith negotiations on nuclear and complete disarmament and to facilitate the development of nuclear energy for peaceful purposes in the non-nuclear-weapon states. The NPT was signed in 1968 and came into force in 1970. In 1995 a review conference will be held to decide how long the treaty will remain in force. Argentina and Brazil are not parties to the NPT which they regard as discriminatory, as unnecessary to ensure their own exclusively peaceful development of nuclear energy and as ineffective in moving the nuclear weapon states toward disarmament.

The treaty's presumed linkage between vertical and horizontal non-proliferation efforts has been largely ignored by the nuclear-weapon states, said a number of panelists who were critical of the nuclear-weapon states' record on nuclear disarmament and on nuclear trade.

Presentations

Martin Gomez Bustillo
Gomez Bustillo pointed out that vertical proliferation has soared in the last two decades. The superpowers now have in their arsenals about 50,000 nuclear warheads, which they have dispersed widely geographically to their allies and which are carried around the world in submarines – a phenomenon that constitutes a third form of proliferation, 'geographical proliferation'. Furthermore, the nuclear-weapon states have made little progress on a comprehensive nuclear test ban, the single most effective measure to restrain vertical proliferation.

Nuclear-weapon states party to the NPT have also failed to keep their bargain to assist the non-nuclear-weapon states party to the treaty in developing the full potential of nuclear energy for peaceful purposes. The unwillingness of the advanced industrial nations to transfer sensitive nuclear technology to non-nuclear-weapon states is a breach of the NPT bargain (a point also made by some earlier panelists). The advanced industrial states have unilaterally restricted transfers of both nuclear and non-nuclear technology, citing the dual-use nature of the technology, even though non-nuclear-weapon states have pledged not to develop nuclear weapons. In the same way the US Nuclear Non-Proliferation Act (NNPA) has prevented exports of technology with potential nuclear weapons applications to non-nuclear-weapon states.

Another example of the treaty's lack of balance is the non-nuclear-weapon states' obligation to accept international safeguards that could compromise their industrial secrets. There is no corresponding safeguards obligation for the nuclear-weapon states, which have the most advanced technologies.

The non-nuclear-weapon states have done a better job of meeting their NPT obligations. The number of nuclear-weapon states has not increased beyond the original five. Although India exploded a peaceful nuclear device in 1974, it has not gone on to develop a nuclear weapon programme. There has been speculation about nuclear weapon programmes in other states but no firm evidence of the development of nuclear weapons.

Any evaluation of the NPT must take note of this serious asymmetry of obligations on the part of the nuclear-weapon and non-nuclear-weapon states. The nuclear-weapon states' greatest contribution to preventing nuclear proliferation would be to agree on a nuclear test ban, thereby halting development of ever-more sophisticated nuclear weapons.

In addition to the NPT, other measures are aimed at preventing horizontal nuclear proliferation, including nuclear weapon-free zones, IAEA safeguards and bilateral agreements such as that between Argentina and Brazil. The 1967 Treaty of Tlatelolco establishes a nuclear weapon-free zone in Latin America. States party to this treaty agree not to test, produce or acquire nuclear weapons, nor to allow any other power to deploy weapons in their territories.

Argentina has not ratified the Treaty of Tlatelolco, although it considers it to be an important disarmament measure. It is concerned that inspections by the Organization for the Prohibition of Nuclear Arms in Latin America (OPANAL), the treaty's regional authority, could compromise its industrial secrets. Argentina also wants the IAEA safeguards required by the Tlatelolco Treaty to recognize PNEs, as permitted by Tlatelolco but not by the NPT. Furthermore, Tlatelolco has no mechanism for verifying that nuclear-weapon states are respecting their commitments to keep nuclear weapons out of the region. Argentina cites the presence of British nuclear weapons in the South Atlantic during the Malvinas War.

As a result of the 1985 Foz de Iguazu agreement, Argentina and Brazil are taking steps on their own to reduce the risk of a regional arms race. They are also working to develop common nuclear policies and strategies to overcome the constraints placed on their nuclear development by their refusal to sign the NPT. In addition, the Foz de Iguazu agreement established a joint working group, 'Legal and Technical Requirements for Safeguards', which has addressed the issue of IAEA safeguards, nuclear trade with third parties and the legal implications of bilateral exchanges of materials and equipment outside the NPT framework.

Today, most industrial nations have the capability to build nuclear weapons but have made a political decision not to do so. It is the political decision to do so that is the key factor. Nuclear proliferation in all its forms – vertical, horizontal, geographical – is to be condemned.

Edmundo Fujita

Fujita criticized the 'non-proliferation ideology' that simplifies and distorts the problem of non-proliferation and diverts attention from its real dimensions. This ideology reflects the interests of those who want to maintain the status quo – not only politically and militarily, but also with respect to science, technology and economics. Its principal characteristics are its exclusive focus on horizontal proliferation, its Manichean approach and its all-inclusiveness. What began as a policy to prevent the horizontal spread of nuclear weapons now extends to nuclear technology, missile technology and chemical weapons.

What are the true dimensions of the non-proliferation problem? In forty-five years the superpowers have gone from zero to 50,000 nuclear weapons. At the same time there has been 'no ascertainable horizontal proliferation, beyond the five declared weapons states'.

It is important to restore the original meaning of nuclear non-proliferation – the linkage between its vertical, horizontal and geographic dimensions – and to address current concerns about vertical proliferation, rather than to chase after prospective proliferators. Fujita proposed the following measures:

- the nuclear-weapon states should take concrete steps to reduce their nuclear arsenals and abandon the theory of nuclear deterrence, which would justify every state's acquiring nuclear weapons.

- pending the complete destruction of their nuclear arsenals, the nuclear-weapon states should offer negative security assurances to the non-nuclear-weapon states.

- the nuclear-weapon states should respect nuclear weapon-free zones and abandon their practice of refusing to disclose whether their ships and submarines are carrying nuclear weapons. These measures must be subject to verification by the non-nuclear-weapon states.

- the nuclear-weapon states should co-operate with non-nuclear-weapon states that want to develop a civilian nuclear industry.

The UN General Assembly's First Special Session on Disarmament identified nuclear disarmament as the international community's highest priority. The non-nuclear-weapon states have kept their part of the bargain; it is time for the nuclear-weapon states to do their part.

Rear Admiral Thomas D. Davies
Davies stated that to understand the issue at hand, it is necessary to review the history of arms control. Arms control agreements have gone from

hortatory expressions of noble principles – the Kellogg-Briand Treaty, for example – to measures such as the INF that have teeth. Today, arms control agreements target a specific class of weapons for elimination. As a result, verification is an important consideration in today's arms control agreements.

With respect to nuclear non-proliferation, the focus must be on ensuring that nations do not have the means to build a nuclear weapon. Safeguards arrangements are part of the verification measures; national safeguards systems are an important component. Ensuring public access to information on national safeguards can be useful in maintaining high national standards.

'Anything nuclear' produces an emotional reaction that sends common sense out the window. This response is true for other areas of high technology as well. Despite their attractiveness, high technology solutions are not always appropriate. A particular kind of wisdom and realism that is hard to come by are needed to deal effectively with these subjects.

Discussion

How meaningful is the NPT? More than 130 nations have chosen to sign the NPT, a US participant pointed out, thereby helping to create an international norm against the spread of nuclear weapons.

Several Argentine participants responded that, to the contrary, widespread adherence to the NPT is less significant than it appears to be. Most NPT states do not have nuclear programmes of any size; therefore, their signature on the NPT is cost free. Furthermore, the value of the Treaty is being eroded by the deployment of nuclear weapons by the superpowers in non-nuclear-weapon states that are parties to the NPT. These states are members of military alliances that rely on nuclear weapons for their defence and have sophisticated nuclear programmes of their own. What is the significance of their signing the Treaty? This 'indirect proliferation' violates the spirit, if not the letter, of the Treaty and is a serious concern.

Only a very few states that are party to the Treaty and that have nuclear energy programmes do not belong to military alliances with nuclear weapons. The NPT should not be credited with preventing the spread of nuclear weapons, nor is nuclear non-proliferation identical with the NPT.

'Latent proliferation' – the stockpiling of nuclear weapon materials that could be quickly turned into weapons – is a serious problem, a US participant said. The US non-proliferation policy should not be 'cost free', as too many authorities in the United States believe non-proliferation is just fine so long as it does not cost the United States anything. However, is it realistic to demand that the United States reduce its nuclear arsenal to zero?

The risk of mass destruction entitles sovereign states to ask another sovereign state to give up its nuclear weapons, an Argentine participant responded. Countries that do not have nuclear weapons have everything to lose if there is a war between the superpowers.

A US participant observed that non-nuclear-weapon states protest that their sovereignty is threatened when horizontal proliferation is discussed, but the same states do not hesitate to tell nuclear-weapon states what to do with their nuclear arsenals. One is no more or less discriminatory than the other. Because these are important issues, 'thoughtful people have a legitimate interest in expressing an opinion about them'.

PANEL SIX. WOULD A BILATERAL ARRANGEMENT BETWEEN ARGENTINA AND BRAZIL SERVE AS A USEFUL MODEL FOR OTHER REGIONS? WHAT WOULD BE THE IMPLICATIONS FOR IAEA SAFEGUARDS AND THE TLATELOLCO AND NPT TREATIES?

Overview

Members of this panel distinguished between two types of non-proliferation systems: one of 'control' and one of 'confidence'. A number of Argentinian and Brazilian participants depicted the NPT safeguards regime as a 'system of control' that has discriminated against Latin America. Argentine-Brazilian cooperation, they said, is based instead on a system of mutual confidence. Some American participants stressed the need for meaningful controls, whether applied bilaterally through a mutually acceptable inspection and verification regime, or internationally by means of IAEA safeguards, or some combination of the two. The panelists and discussants reviewed the advantages and drawbacks of various systems of control, debating the need for safeguards on weapons-usable nuclear material.

Presentations

Minister Roberto García Moritán
García Moritán criticized the international non-proliferation regime for restricting trade in sensitive nuclear equipment and technologies, ostensibly to prevent nuclear proliferation. This approach is perpetuating the commercial advantage of a few nations. By focusing on a strategy of 'controls', the non-proliferation regime has failed to recognize the importance of

'confidence'. Non-proliferation is a political issue that requires a political solution. Although the process that Argentina and Brazil have followed cannot be a 'model' for other regions, it can provide some useful insights into policy alternatives for non-proliferation.

Argentina has followed the path of the industrialized nations in seeking the benefits of nuclear energy. It was the restrictive trade policies of the industrialized nations, intent on maintaining their commercial advantage, that forced Argentina to develop complete fuel cycle facilities. Argentina had to invest its scarce resources to develop an indigenous enrichment technology in order to assure a reliable supply of fuel for its powerplants and for the research reactors it exports. Argentina's decision to proceed with enrichment technology followed a US cutoff in the supply of fuel for its research reactors.[7]

An international consensus must be developed that allows all nations unlimited access to nuclear technology for peaceful nuclear programmes, based on their interests, needs and priorities. He criticized the 'IAEA/NPT/ London Club/fullscope safeguards regime' as discriminatory, favouring European nations and Japan over the countries of Latin America. He cited the IAEA's refusal to accept the Treaty of Tlatelolco as a basis for safeguards agreements and the Agency's replacement of item-specific safeguards with the NPT's full-scope requirements. The Treaty of Tlatelolco, he said, has suffered unfairly because of the controversy over peaceful nuclear explosions (PNEs) and the refusal of traditional suppliers to recognize it as a guarantor of peaceful use and therefore as a basis for exports. Under these circumstances, it is not surprising that some Latin American nations have chosen to remain outside the NPT regime.

What are the implications of the bilateral agreement between Argentina and Brazil for the NPT regime? By eliminating competition and promoting cooperation, the agreement between Argentina and Brazil has far outpaced the achievements of the NPT system. Could there be an accommodation between the NPT regime and the Argentine-Brazilian accord? It is difficult to predict how the Argentine-Brazilian process will evolve, but the inflexible attitude of the IAEA with respect to the Tlatelolco safeguards and the difficulties the Europeans have faced in negotiating EURATOM safeguards that are compatible with the NPT safeguards regime leave little room for optimism.

Could this process serve as a model for other regions? Each region is unique and therefore the experiences of Argentina and Brazil may not be applicable elsewhere. However, international acceptance of the two nations' nuclear agreement could stimulate similar processes elsewhere, as part of an effort to get beyond the 'rigid schemes' of the past.

An analysis of Argentina's and Brazil's experience leads to the following recommendations:

1. Broaden the inventory of possible approaches to non-proliferation beyond the NPT, because the Treaty interferes with a nation's right to determine its own technological development.
2. Consider confidence-building mechanisms in addition to safeguard controls. This recommendation is not a criticism of safeguards, which remain an important mechanism and which are appropriate in certain circumstances.
3. Devise political solutions to problems that are fundamentally political. These solutions must be based on an international consensus that takes into account the needs of *all* nations and recognizes that the era of postwar confrontation and suspicion has ended.

Counsellor José Felicio
Felicio restated that Argentina and Brazil are developing a system based on cooperation, not controls. 'If the need arises to bring certain aspects of our nuclear cooperation under controls, I am certain both governments will turn to the IAEA to ask for appropriate inspections. That may happen in the future. It has not happened so far.'

Events overtook the joint working group's discussions about establishing a system of 'controls'. Technical and political contacts between the two nations established a degree of mutual confidence through cooperation that made formal inspections unnecessary. Felicio cited the exchange of visits to sensitive nuclear facilities and a joint breeder reactor research project as evidence of this process.

Furthermore, a bilateral system of formal inspections between Argentina and Brazil would undermine the IAEA and affect its credibility. Nuclear cooperation is the best way to control nuclear proliferation. Addressing the chair Felicio asked, 'What kind of arrangement would replace this intimate arrangement?'

Dr Victor Gilinsky
Gilinsky acknowledged that 'whether informal arrangements are superior to formal inspections' is ultimately for Argentina and Brazil to decide. However, their decision will influence other countries that are considering participation in an international safeguards regime. An appropriate analogy is the baggage inspection system used by airlines to inspect luggage as passengers are boarding. Suppose that some passengers declined to allow their luggage to be inspected and explained they had a prior

agreement 'just to trust each other'. What would the other passengers think?

The safeguards system is important because it erects a barrier between commercial and military use of nuclear materials. Critics of the inspection system are mistaken in arguing that it is too intrusive. In fact, 'the system is too slow. . . . That is why it is dangerous to have unhampered use, worldwide, of . . . plutonium and highly enriched uranium'.

Although he urged restraint in the commercial use of nuclear-weapons materials, 'Every country has a right to pursue its development as it sees fit. But, . . . if everyone pushes to the legal limits of whatever sovereignty will permit, I think we will get ourselves into a lot of trouble on a worldwide basis.'

Ambassador Hector Gros Espiell

Gros Espiell praised the agreements between Argentina and Brazil. Bilateral agreements between developing countries that are not parties to the NPT are one way to promote peaceful uses of nuclear energy. They can serve as an instructive example to other countries at a similar level of economic and technological development.

Because these are agreements to promote the *peaceful* use of nuclear energy, it will be necessary at some future date to submit them to an international safeguards regime so that they are beyond any suspicion. No one has suggested they should be beyond any type of control. Rather, it is a question of finding the appropriate system of controls, a system that can be freely accepted and not imposed – a system that is compatible with the demands of national sovereignty.

International safeguards today are, unfortunately, based on the NPT. However, a rejection of NPT safeguards should not be seen as a repudiation 'of the very idea of international safeguards'. Safeguards based on the Treaty of Tlatelolco could co-exist in Latin America with other kinds of safeguards arrangements for nations that are not parties to the Tlatelolco Treaty. This objective is not likely to be realized in the short term. Meanwhile, Latin American solidarity is a fundamentally important objective in itself.

For a 'universal' norm, one should look to the IAEA, not to the NPT. The IAEA should not be merely an instrument for applying the NPT. If the *objective* of safeguards is kept in mind there can be greater flexibility in negotiating safeguards agreements so as to take regional factors into account.

The agreements between Argentina and Brazil should be of capital importance for other Latin American nations, whether they are parties to the NPT, Tlatelolco or both. Argentina and Brazil are parties to neither the NPT nor Tlatelolco. Therefore, the agreements of cooperation between them represent a 'valid effort of the highest interest' to all nations, but especially to those of Latin America.

Discussion

The evolving system of nuclear cooperation between Argentina and Brazil does not disregard the international non-proliferation regime, an Argentinian commented. Although it has no safeguards requirement, the overall agreement does provide for 'mutual consultations' with respect to the eventual application of international safeguards on bilateral nuclear transfers, and it requires a peaceful use commitment.

This understanding was confirmed by a Brazilian, who said that the question of safeguards had not yet been decided. However, one of the three working groups established by the agreement deals precisely with the relationship between Argentina and Brazil and the overall non-proliferation system.

Another Brazilian spoke in support of 'mutual trust' as a substitute for safeguards.

The American chairman, responding to Felicio's earlier question about what kind of arrangement could replace one of 'intimate' cooperation, posed a hypothetical situation in which, in a relationship based solely on mutual trust, one side discovers that the other side is producing and accumulating substantial quantities of weapons-usable materials without any provisions for public accounting or verification of peaceful use. How would a mutual trust arrangement respond to such a strain?

A Brazilian replied that although mutual trust might not be a sufficient basis for an arms reduction agreement, Argentina and Brazil are agreeing to co-operate in the peaceful use of nuclear energy, an entirely different matter. In any case, it is the responsibility of the Brazilian Congress to determine the disposition of nuclear facilities and nuclear materials in Brazil, under the terms of Brazil's new constitution.

An Argentinian spoke in favour of Argentina's and Brazil's applying safeguards to fissionable material. A safeguards agreement would strengthen the IAEA and pave the way for the participation of other Latin American nations. It would demonstrate that this arrangement is not just a joint venture of Argentina and Brazil. However, a full-scope safeguards arrangement is not being considered, according to Argentine participants. As long as Argentina is denied access to nuclear fuel cycle technology, it will not accept full-scope safeguards. We are prepared to 'exchange technology for safeguards, the guarantee of non-proliferation for international cooperation'. But we are not prepared to pay up front without a contract for delivery.

There is no one system of 'safeguards', an Argentinian pointed out. There are IAEA item-specific safeguards (INFCIRC 66/Rev. 2), NPT full-scope safeguards (INFCIRC 153), EURATOM safeguards and a somewhat different safeguards arrangement negotiated by Japan. Argentina and Brazil want a system of safeguards based on the Tlatelolco Treaty.

The IAEA was praised by both Argentinian and Brazilian participants, who made a distinction between the IAEA and the NPT. The IAEA is a 'distinguished organisation' that has provided significant technical assistance to the nuclear programmes of less developed countries. If Brazil and Argentina were to adopt safeguards, they would turn to the IAEA, a Brazilian said, rather than attempt to develop a bilateral system that would undercut the IAEA's credibility.

The NPT system, on the other hand, has had an adverse impact on the development of nuclear energy, Argentinian participants said. Potential benefits to Argentina of subscribing to the NPT were eliminated when the London Club was organized, because it blocked exports of sensitive fuel cycle technology to NPT parties as well as non-parties.

Brazil will not sign the NPT, a Brazilian said, because the country's military and foreign ministry experts and intellectuals see the real objective of NPT advocates as being 'to restrain Brazil's technological development and therefore its economic growth'. Brazilians are suspicious of the close ties between the US administration and private US economic interests that would benefit from curbs on Brazilian economic activity.

An Uruguayan participant noted that Uruguay, as a party to the Nuclear Non-Proliferation and Tlatelolco treaties, and as a subscriber to full-scope safeguards, would qualify to receive transfers of nuclear technology under Article IV of the Treaty that could not be exported to Argentina and Brazil as non-NPT states, but that might be used for the benefit of the region. The American chairman responded by raising an idea he had discussed informally outside the conference room . . . namely, the construction in Uruguay of a nuclear powerplant to supply power to Argentina, Brazil and Uruguay as part of economic integration in the region. Nuclear supplier states whose laws or policies prevented exports to Argentina and Brazil would face no such obstacles in the case of Uruguay, he pointed out.

Two undercurrents run through these discussions, an Argentine observed. They relate to technological autonomy and national sovereignty – in other words, to 'power' – and to whether Argentina and Brazil can be relied upon to exercise power in a responsible way. Argentina and Brazil, 'like any country defending its national interest', will not accept safeguards on the products they develop, the Argentine continued. With respect to their 'reliability', the unpredictability of previous regimes may have raised questions. The best guarantor of rational, responsible and reliable behaviour is the consolidation of democratic regimes in Argentina and Brazil. With respect to 'reliability' in its nuclear exports, Argentina has shown 'the highest degree of reliability' by requiring safeguards as a condition of export.

To ensure the survival of cooperation between Argentina and Brazil, some controls are necessary. It was the speaker's understanding that safeguards are in fact under discussion in one of the joint working groups. That no one is prepared to acknowledge this on the record probably indicates only the absence of agreement thus far.

In his concluding remarks, a Brazilian stressed the importance of a national system of controls. Greater 'transparency' in Brazil's domestic nuclear programme was a prerequisite for a meaningful bilateral agreement, and therefore, there should be controls on nuclear power and nuclear submarine programmes. Although Brazil's new constitution gives Congress responsibility for oversight of nuclear programmes, he said, these congressional powers exist 'just on paper'. The Brazilian Congress does not yet have the expertise to review the budget in a meaningful way, and it may take the next two presidential terms for it to implement its constitutional powers.

Another Brazilian noted in response that the budget committee was invited to conduct an on-site inspection of nuclear projects in Brazil before approving the budget.

The chairman, in response to an objection that these topics were out of order in a conference on bilateral relations, stated that discussion of domestic issues is appropriate on the premise that an adequate national system is a prerequisite for an effective bilateral regime.

Another US participant, citing earlier discussion of the need for a safeguards system tailored to Argentina's requirements, asked how such an Argentine system would differ from those in place in NPT nations. An Argentine replied that while he had 'no specific ideas', Argentina would not accept NPT-type safeguards on the full scope of its nuclear activities.

The conference chairman, in closing remarks, emphasized the need to put a meaningful safeguards system in place before either nation has produced significant quantities of weapons-usable material. The acquisition of only one or two nuclear weapons when the other side has none would be of tremendous military significance.

One desirable outcome of the *rapprochement*, he suggested, would be a verifiable agreement to produce *no* weapon-usable nuclear material. If fissionable material is going to be produced, however, there should be a system to keep track of how much is produced and where it is to permit both domestic and bilateral accountability. 'It is a matter of enlightened self-interest that also would have great regional and global benefits', the chairman said. Although there now appears to be no visible movement toward formal bilateral safeguards, he noted that Argentine and Brazilian participants had indicated a safeguards working group exists.

The conference also made clear, he suggested, the linkage between vertical and horizontal non-proliferation efforts – namely, that the United States and the USSR must meet their Article VI commitments to make their non-proliferation objectives credible. How can non-nuclear weapons states reciprocate? Perhaps the time has come formally to renounce PNEs. There also could be reciprocity if a comprehensive test ban agreed to by the superpowers were followed by a universal cut-off in the production of weapons-usable materials. If nuclear weapons are not to proliferate, and if the nuclear arms race is truly to be halted, an international norm must be established that holds *both* nuclear explosions and weapons-usable materials to be illegitimate, either for peaceful or military purposes.

The chairman concluded that much 'food for thought' had been placed on the table at this meeting. On behalf of all the participants, an Argentine thanked the chairman and the Nuclear Control Institute for organizing this 'very stimulating' meeting.

2 Goals of Argentine-Brazilian Nuclear Cooperation

Presentation: *Ambassador Julio C. Carasales*[1]

OVERCOMING THE PAST

> Argentine-Brazilian relations at present concentrate the greatest efforts
> of the political imagination and of the creative force of both govern-
> ments and are beginning to attract the attention of the business, sindical
> and cultural spheres on both sides of their common frontier. This
> process, which up to the present is reflected by 22 existing Protocols
> of cooperation and integration, represents a tangible reversal of the
> historical antagonism and distrust of the past.[2]

The preceding statement describes, briefly but accurately, the most
important political-economic phenomenon in South America in the recent
past: the initiation of a process of real and effective integration between
the two most powerful countries of the continent, Argentina and Brazil. If
this process is definitively consolidated and a common market is created
– the established goal of both countries – it will no doubt be, at least in
the opinion of the author, the most important event in this region in this
century.

Perhaps the international community has not yet clearly perceived
the extraordinary scope of this development and all its implications. It
is probably premature to make concluding judgements on this matter,
and preferably only cautious indications of hope should be made. Past
experience also weighs heavily in an evaluation of this Argentine-Brazilian
process: it is characterized by more than a century and a half of conflict and
rivalry and by the failure of some efforts at *rapprochement* in the face of
the disappearance of the political figures that promoted them.

Both countries became independent simultaneously at the beginning of
the last century. From the start they perpetuated the adversarial relations
of the Spanish and Portuguese empires in South America, especially in the

region of the River Plate (Rio de la Plata). The Argentine Confederation and the Brazilian Empire were at war from 1825 to 1826, and general mistrust, competition and antagonism characterized relations between the two from then on, with occasional episodes of cooperation, inevitable given their lengthy and inescapable political, economic, geographic and geopolitical links. Nobody can deny the fact that for many years the military in both countries always considered as a priority in their 'hypotheses of conflict' the possibility of a conflict with the neighbouring country.

Beyond impressive but unrealistic statements such as that of President Rogue Sáenz Peña of Argentina[3] – 'Everything unites us and nothing separates us' – it was not until the beginning of the decade of the 1950s, during the governments of Presidents Juan Peron and Julio Vargas, that the first serious attempt at extensive cooperation between the two countries was made. The effort proved futile, probably because 'it consisted of a precarious approach related to internal political projects that could not count on a solid support on the diplomatic and economic levels'.[4]

The second instance of *rapprochement* occurred during the presidency of Arturo Frondizi of Argentina and his counterparts in Brazil, Presidents Juscelino Kubischek and Janio Quadros. The main resulting documents were issued at the Conference of Uruguaiana on 21 April 1961. This effort again proved fruitless, as one year later both President Frondizi and President Quadros were deposed. Nevertheless, as noted, 'undoubtedly some favorable consequences remain from each successful step. In this case the possibility of war between both countries is almost negligible and the need for cooperation, despite its transitory eclipses, has proved to be meaningful and has become a center of attention for many Brazilians as well as Argentines'.[5]

In support of the above point of view, it is sufficient to remember the statement by the future foreign minister of Brazil, Antonio Azeredo de Silveira, who in 1969 said,

> The integration of Brazil and Argentina is a prospect that presents almost ideal conditions for its consolidation. It tends to be a natural phenomena and should occur globally. Obviously, it can be hastened or slowed down according to the degree to which both Governments acknowledge that in this lies the key to the solution of many of the issues that condition the economic and social structures of both countries.[6]

One problem delayed acknowledgement of the increasing need for a new effort at *rapprochement*: the controversy over the use of shared natural resources, specifically the rivers in the Rio de la Plata Basin, especially the

Paraná, for hydro-electricity. Only in 1979, after an arduous and difficult negotiation, was the issue resolved satisfactorily. The road was then open for initiating a new phase. In 1980 Argentina and Brazil entered into a number of agreements and conventions in different areas. Perhaps the most important, because of the nature of the issue, was that on nuclear cooperation (referred to later).

In the context of this brief overview of some of the most important milestones in Argentine-Brazilian relations within the last decades, it can be said that the *rapprochement* of 1980 did not have the depth or the consequences many desired. The reason probably was that the internal situation in both countries – they were moving from military to civilian governments – was not the most favourable for initiating Argentine-Brazilian integration.

Adequate conditions emerged only when, in both countries, elected democratic presidents assumed power, Raul Alfonsín in Argentina in December 1983 and José Sarney in Brazil in March 1985, succeeding deceased President-elect Tancredo Neves. It appears that the popular mandate of both these presidents supported them in undertaking important projects, not the least of which was the initiation of a new era in Argentine-Brazilian relations.

On 30 November 1985 both presidents met in Foz de Iguazú. They launched a programme that was ultimately formalized with the signing of a historically important document in Buenos Aires on 31 July 1986, the Act for Argentine-Brazilian Integration, accompanied by 12 protocols of cooperation covering different topics, including nuclear cooperation. New protocols were signed at subsequent presidential meetings, reaching a total of 24 as of the end of 1989.

On his recent visit to Brazil (22–24 August 1989), the new president of Argentina, Carlos Menem, emphasized his firm conviction to support the integration already initiated. He signed new protocols, dispelling any doubt about the degree to which the new Argentine government would back the efforts initiated under the previous government, now the opposition. His stand is clear and proves that, at least on the part of Argentina, *rapprochement* with Brazil is not a passing fancy or the product of favourable circumstances. To the contrary, it is the outgrowth of deep convictions and a perception that meeting the challenges of the world today requires overcoming provincialism and integrating extensive geographic areas, starting with a dynamic neighbouring nation rich in natural resources. The ultimate goal is wider integration that would progressively encompass all Latin American nations. This goal has been furthered by the incorporation of the Republic of Uruguay into the integration process.

The above-mentioned advances do not mean that the road has been easy or free of pitfalls, difficulties and setbacks, as noted. Moreover, it is wrong to ignore existing problems, especially in the economic and commercial spheres; while not addressed in this chapter, overcoming them will require energetic and imaginative measures. At the same time, it is important not to lose sight of the fact that, little by little, the integration of two great South American countries is becoming irreversible. It may not yet be complete, but it will be.

It is difficult at present to evaluate adequately the extent and importance of this integration. A page in history has been turned, at least in the history of South America. With Argentina and Brazil intimately united, and no longer separated by resentment, disagreements and rivalries, but, on the contrary, politically and economically co-ordinated to face the rest of the world – a source of stimuli, challenges and difficulties – it seems obvious that the future of this part of the world will follow a path different from the traditional one.

No one should fear that this new reality will generate threats or inconveniences to neighbouring or distant countries. On the contrary, the opposite is more likely. The benefits of this Argentine-Brazilian *rapprochement* are obviously not restricted to South America. South America is itself a privileged continent, and although it is facing the drama of economic and social underdevelopment, it need not be concerned – at least since World War II – with the risk of armed conflicts between nations of the region. This risk would disappear entirely with the consolidation of a Brazilian-Argentine agreement. Further, the strengthening of the economies of these two countries as a result of their policies of integration and the eventual creation of a common market can only benefit the other countries in the region.

The strengthening of South America as an area of peace and progress will benefit not just the rest of Latin America, but also the world. It is not unrealistic to characterize the integration of Argentina and Brazil as the establishment of a binational nucleus based on peace, security and development that could favourably influence even geographically remote areas. It must not be forgotten that security is indivisible and that no conflict is purely local, without effects on apparently unrelated areas.

It is within the framework of this historically important effort initiated by Argentina and Brazil that the policy of cooperation in the peaceful use of nuclear energy undertaken by both countries can be established, together with a similar effort in other fields and subjects. It is easy to understand that because of the nature of the subject and the implications for national defence and security, cooperation in the nuclear field cannot be undertaken in an isolated manner, that is, in the absence of a deep and general desire for cooperation and integration. In the atmosphere of mistrust and reciprocal fear that prevailed historically, cooperation on nuclear matters was unthinkable. This situation no longer pertains, as facts have proven. In the following section of this paper, I address the main events of this joint effort. It undoubtedly is of enormous significance in itself and as an indicator that the international suspicions and concerns that have been voiced repeatedly have less and less reason to exist.

Outside the nuclear field itself, it must be noted that this particular effort of cooperation 'represents the main political triumph of the Argentine-Brazilian process of integration. The political impact of the nuclear *rapprochement* between Argentina and Brazil became, to a large extent, the counterpart of the difficulties encountered in the economic field'.[7]

NUCLEAR COOPERATION

Argentina and Brazil, technologically the most advanced countries in Latin America, logically have pursued energy produced by atomic fission. They chose different paths, however, Argentina using natural uranium and Brazil enriched uranium. Further, they initiated and developed their processes at different rates. They did, however, coincide in the quest to establish complete control over the production cycle of the fuel and in their rejection of discriminatory external controls that might restrict their independent development of nuclear energy for peaceful purposes in the most convenient manner.

Until 1980, a non-institutionalized form of cooperation existed between the Argentine National Commission of Atomic Energy (Comisión Nacional de Energía Atomica, CNEA) and different research groups in Brazil, mainly academic institutions. However, the lack of interest on the part of the political and nuclear authorities in the two countries in formalizing this informal cooperation and extending it to other areas was notorious.[8]

The foreign policies of both countries on nuclear issues were quite similar and, on many occasions, parallel. However, prior cooperation rarely existed, nor was the prevailing atmosphere favourable to strengthening their common goals. The long history of mistrust and rivalry prevailed, particularly in the case of sensitive matters such as nuclear issues. Another factor complicated the situation: Argentina was ahead of Brazil in nuclear development in general, a serious obstacle to the creation of any type of association.

These unfavourable circumstances changed toward the end of the 1970s, as pointed out. In 1979 the important dispute that separated both countries with respect to the use of their shared natural resources was resolved, the last in a long series of misunderstandings and misgivings that traditionally had characterized their relations. Moreover, in the nuclear field specifically, Brazil believed that it had reached a similar level of development as Argentina had, so that a discussion between equals was now possible.

The visit of the president of Brazil, General João Figuereido, to Buenos Aires between 14 May and 17 May 1980 produced two indications of a new spirit in the nuclear field. The foreign ministers of both countries signed an 'agreement of cooperation for the development and application of the use of nuclear energy for peaceful purposes' and both presidents issued a joint declaration on 17 May that included an important paragraph on nuclear policies. The agreement mentioned the areas in which cooperation would be developed and the means by which it would be undertaken (scientific and technical training of personnel, exchange of experts, professors and students, reciprocal consultations, mutual provision of equipment, materials and services, exchange of information, and so on). The presidential declaration expressed 'particular satisfaction with the documents signed on matters of nuclear cooperation' and *'ratifies emphatically,* as well, that the nuclear programmes of both countries *pursue peaceful purposes exclusively* and that they *oppose the development of atomic weapons'* (emphasis added).

The agreement of 17 May 1980 was a framework-treaty used by the competent technical organizations for signing conventions and protocols relative to the specific aspects contemplated in the agreement. On the same day, agreements of cooperation on basic research, the production of radioisotopes and marked molecules, radiological protection, nuclear security and so forth were also signed by the Argentine National Commission of Atomic Energy and the National Commission for Nuclear Energy of Brazil.

In accordance with the agreement, months later, on 20 August 1980, in Brasilia, two documents were signed: the Protocol of Execution No. 1 on cooperation in the training of personnel in the nuclear sector, which in practice worked unsatisfactorily; and a Protocol of Execution No. 2 on technical nuclear information, which to the contrary worked acceptably well.

Also on 17 May 1980 another convention on cooperation was signed by the CNEA and Empresas Nucleares Brasilenas, S.A. In turn it produced a protocol on industrial cooperation, permitting contracting with commercial entities for the loan and consumption of uranium concentrates, for the provision of zircalloy tubes, for the manufacture of the lower section of the lower area of the pressure receptor for the Argentine nuclear power plant of Atucha II, and so on.

It is not within the scope of this paper to describe all the events relating to Argentine-Brazilian nuclear cooperation. However, a review of what has occurred since 1980 reveals that implementation of the agreement signed that year, even though irregular and not to be underestimated, did not satisfy the awakened expectations of the signing parties. Political and economic difficulties in both countries distracted the attention of both governments, which were occupied with managing the complicated transition from military regime to civilian government. These circumstances explain in part why the great political decision to initiate this fundamental change in relations between both countries proved somewhat ineffective, even though it seemed timely and was certainly intended to include nuclear cooperation.

The democratic governments of President Raul Alfonsín, who took office on 10 December 1983, and of President José Sarney, who took office little more than a year later, would be responsible for taking that important step. As noted, the presidents, meeting on 30 November 1985 in Foz de Iguazú, signed a number of important documents, among them the Joint Declaration on Nuclear Policy.

The joint declaration contained several elements that should be emphasized: (1) it reiterated 'the commitment to develop nuclear energy *for peaceful purposes exclusively*' (emphasis added); (2) it stated the goal of 'closely cooperating in all fields of peaceful use of nuclear energy and to complement each other in those aspects deemed convenient'; (3) it expressed 'the hope that this cooperation be extended to those other Latin American countries that had similar goals'; and (4) it averred that the cooperation between both countries not only 'constitutes a multiplying factor of the beneficial effect that the use of nuclear energy can provide both countries' but also 'allows both countries to better confront the increasing difficulties found in the international supply of equipment and materials'.

As can be concluded from reading the joint declaration, both presidents did not limit themselves just to broad statements. Much to the contrary, they put into practice the high goals proclaimed in the document by establishing 'a joint working group *under the responsibility of the Argentine and Brazilian Foreign Ministries* with the participation of representatives of the respective Commissions and nuclear enterprises' (emphasis added). It is worth pointing out that, among the goals of the working group, was 'the creation of mechanisms that could insure the higher interests of *peace, security* and the development of the region' (emphasis added).

It is probable that, in recognition of the importance of the political factor in carrying out this process, it was decided that the working group would operate within the jurisdiction of both foreign ministries, whereas 'the technical aspects of nuclear cooperation' . . . 'would continue to be governed by already existing instruments', that is, those established in 1980 and the years following.

The working group on nuclear policy has become the co-ordinating and promoting force for Argentine-Brazilian cooperation in the nuclear field. Following the Declaration of Iperó (8 April 1988) the working group became a permanent committee, with a mandate to meet every 120 days in each country alternatively (this practice was already being followed).

The working group, later the permanent committee, channelled its work through three sub-groups: *technical cooperation; coordination of foreign policy;* and *legal and technical requirements of cooperation.*

Simultaneously, Argentine and Brazilian companies in the nuclear sector have also been meeting to promote the integration of their respective industries. To that end they created the Argentine-Brazilian Nuclear Sector Business Committee (Comité Empresarial Argentino-Brasileno del Area Nuclear, or CEABAN).

THE POLITICAL EVOLUTION OF NUCLEAR INTEGRATION

The process of cooperation and nuclear integration between Argentina and Brazil has evolved basically at two levels, the political and the technological-industrial. Given the nature of this presentation, I refer here specifically to the first.

The series of visits and joint declarations by the chiefs of state can be considered the milestones of this integration, together with the signing of important bilateral agreements. The following is a list of 'joint declarations on nuclear policy':

1. Declaration of Foz de Iguazú, 30 November 1985
2. Declaration of Brasilia, 10 December 1986
3. Declaration of Viedma, 17 July 1987
4. Declaration of Iperó, 8 April 1988
5. Declaration of Ezeiza, 29 November 1988.

Beyond the content of these joint declarations, to which I refer later, it is truly significant that each was issued in the course of an official visit of the president of one of the countries to the other. If the visit by the president of Brazil to Buenos Aires in May 1986 and to Brazil by the new Argentine president in August 1989 were included, it would be difficult to find more frequent contact between the heads of state of two countries or a clearer indication of decisiveness and intensity in the pursuit of integration.

Three of the visits had an unusual component: a trip by the president of Brazil in July 1987 to the uranium enrichment plant in Pilcaniyeu, Argentina; a trip by the President of Argentina in April 1988 to a similar plant at the Aramar Experimental Center in Iperó, Brazil; and a trip by President Sarney in November 1988 to the laboratory of radiochemical processes of CNEA in Ezeiza, Argentina.

As was pointed out, the important Act of Integration of 31 July 1986 has been the source of numerous cooperation protocols in different sectors, which continue to be signed as integration advances. Together with the Integration Act, Protocol No. 11, Immediate Information and Reciprocal Assistance in Case of Nucléar Accidents and Radiological Emergencies, was signed prior to the conventions on the same subject promoted by the International Atomic Energy Agency (IAEA), to which two annexes that were agreed to in Brasilia on 10 December 1986 can be added. It must be noted that the system of assistance worked efficiently in the case of the accident at Goiania in Brazil.

On 10 December 1986 Protocol No. 17 on Nuclear Cooperation was signed in Brasilia. It defined several areas in which mutual cooperation and development could be achieved: (1) high density fuels for research reactors; (2) detectors, electronics and nuclear instrumentation; (3) enrichment of stable isotopes; (4) research on nuclear physics and plasma physics; (5) safeguard techniques in light of the commitments made by both parties with the IAEA; (6) in the long term, technical and economic viability of joint development of a demonstration fast breeder reactor; and (7) non-destructive techniques of assays of materials used in nuclear technology.

Finally, on 23 August 1989, in Brasilia, the foreign ministers of both countries signed an agreement 'to promote the extensive industrial complementation to the nuclear sector' (Annex I to Protocol No. 17).

Analysis of the documents signed as of the first agreement on 17 May 1980 reveals a series of factors that underlie the close cooperation on nuclear matters between Argentina and Brazil. The following are worth pointing out:

(1) *Reaffirmation of the peaceful purpose of Argentine-Brazilian nuclear programmes.* In the joint declaration of 17 May 1980, as well as in the declarations of Foz de Iguazú, Viedma, Iperó and Ezeiza, the chiefs of state emphatically and solemnly reiterated the peaceful

purpose for the development of nuclear programmes in both countries. *This commitment expressed in successive international documents cannot be ignored or undervalued.*

(2)   *Strengthening mutual confidence.* This subject, which is always present, was expressly mentioned in the Declarations of Brasilia, Iperó and Ezeiza. The Declaration of Iperó, for example, pointed out the fact that

> bilateral cooperation in the nuclear field introduces unprecedented forms of cooperation promoting an increasing number of visits, political and technical contacts, and significant exchange of information, contribution, in its entirety, to the *consolidation of the mutual confidence.*

It further expressed a wish to '*deepen the reciprocal confidence*' (emphasis added). The Declaration of Ezeiza addressed the issue of the 'consolidation of the atmosphere of *mutual confidence* reached as a result of the increasing and continual reciprocal exchange of knowledge and of the joint effort in the execution of important projects, *an event without precedent between two developing countries in the same region*' (emphasis added).

(3)   *Use of the advances derived from the peaceful use of nuclear energy for the benefit of the people of both nations.* After the Declaration of Foz de Iguazú, this goal appears to have become central to the building of nuclear cooperation, given that both countries are making a huge effort to develop their technical capabilities in this field and intense cooperation in this field would widen this capacity and have a multiplier effect on the better utilization of nuclear energy to promote the economic and social well-being of the people of both nations.

(4)   *Possibility of extending cooperation in nuclear matters to other countries in Latin America.* Since the Declaration of Foz de Iguazú, reiterated by the Declarations of Iperó and Ezeiza, there is, as mentioned in this last declaration, 'the permanent will to extend the cooperation and interchange of nuclear technology to all Latin American nations interested in having access to it'.

(5)   *Coordination of common foreign policies on nuclear matters.* Article IX of the Cooperation Agreement of 17 May 1980 already contemplated consultations to co-ordinate positions. The Declaration of Brasilia mentioned strengthening of 'the coordination of political positions in the international sphere, for the defense of common interests', while those of Iperó and Ezeiza emphasized 'the complete agreement of the Argentine and Brazilian positions on major international issues of nuclear energy', among which the prohibitions and hindrances that affect international cooperation with respect to nuclear supplies are worth mentioning. In the Declaration of

Iperó, both governments 'reaffirm the inalienable right to develop, *without restrictions, their nuclear programs for peaceful purposes*' (emphasis added).

(6)   *Concern for the peace and security of the region.* The important Declaration of Foz de Iguazú referred to 'the creation of mechanisms that insure the superior interests of peace, security and the development of the region', a concept again addressed in the Declaration of Brasilia, which stressed the 'determination to reinforce the necessary mechanisms in order that nuclear science and technology of both countries act as effective reaffirming factors in the interests of peace, security, and development' and referred to the need for coordination 'to preserve the region from the *risk of the introduction of nuclear weapons*' (emphasis added).

FINAL CONSIDERATIONS

The process of cooperation and nuclear integration between Argentina and Brazil, summarized in this paper, merits, in my opinion, a few final comments, as follows.

(1)   *Uniqueness of the process.* Even though mentioned several times, it is not redundant to emphasize the unique character, at least among developing countries, of this effort at *rapprochement* on such a delicate subject as that of nuclear issues, particularly so as it involves countries that traditionally have been rivals. For those who do not belong to this Latin American region, it may be difficult to understand the full implications of this joint enterprise, of which the nuclear sector represents only part of a wider and varied spectrum that joins in common operation the policies and actions of the two largest and most advanced countries of South America.

In this respect, despite the differences and distances, this event can be compared with the '*rapprochement*' of France and the Federal Republic of Germany after World War II, which led to the establishment of the Coal and Steel Community and later of the European Common Market which has radically changed the economic, social and, soon, political map of Europe. Argentine-Brazilian integration is now in its first steps, but the road it has to travel is not that different from the European one. In the future it too may lead to not very different accomplishments.

We must not forget that there is a definite desire to open this process to other countries in Latin America, achieved to some degree in the case of Uruguay.

(2)   *Continuity of the process.* During the last few years, as the policies of nuclear cooperation between both countries have been strengthened and

consolidated, an increasing concern in the international arena has been the following: would this process survive the changes in governments in Brazil and Argentina about to occur in 1989? In the case of Argentina, it seemed increasingly possible that the Justicialist Party would win the elections. In that case an abundance of nationalistic attitudes toward nuclear policies were expected.[9]

The facts quickly dissipated the concerns. The newly appointed president, Carlos Menem, who took office 8 July 1989, made his first trip abroad to Brazil, one and a half months later. Both countries' foreign ministers signed an important agreement on industrial complementation that began by reaffirming all previous agreements and joint declarations promulgated since 1980. When signing this and other documents, among which was an important Joint Declaration on Space Cooperation, the Argentine chief of state made a speech in which he stated,

We feel proud of the process initiated with the Declaration of Iguazú of 1985, that permitted consolidating reciprocal trust, the exchange of experiences, the sharing of technologies and the generating of new research. We jointly defend our inalienable rights of reaching the next century with a shared and individual scientific and technological basis for the economic and social benefit of our people (emphasis added).

The presidential elections in Brazil will be held on 15 November 1989. Even though in politics no one can make predictions, nothing leads me to believe that the policy of cooperation with Argentina might change with a simple transfer in government. What is at stake in this process is too great to be affected by changes in people or parties. If the policy of integration experiences a fundamental change in the future, it will be the result of more important causes from which no country is exempt and which cannot be predicted at present.

(3)   *Future evolution of this process.* In a task so great and comprehensive as that which Brazil and Argentina have undertaken, difficulties and inconveniences are unavoidable, and progress cannot occur at the same rate in all sectors. Even within one sector the rate of progress varies over time or even becomes stagnant.

The situation described above certainly applies to cooperation on nuclear issues. It is undeniable that the process initiated in 1980 and intensified in 1985 has progressed considerably and, as was noted, has become something unique, at least among countries with the characteristics of Brazil and Argentina. While it is pure speculation to predict how far this process might go, it is undeniable that the possibilities are great and promising.

There might be moments of relative non-productivity, sometimes necessary to permit digesting and consolidating what has already been achieved, but a tendency to regress seems very unlikely.

A point made frequently, especially internationally, in connection with Argentine-Brazilian nuclear cooperation, relates to the creation of a system of mutual 'inspection' or 'verification' that would guarantee one country that the other was not working to produce nuclear weapons. Security concerns are natural and legitimate for any country, and obviously Argentina and Brazil have them. It seems, however, that a belief exists that these concerns can only be taken care of by a system of inspections and reciprocal controls that probably includes some type of safeguards.

Although no possibility can be excluded *a priori*, even those mentioned below, it needs to be pointed out that the words 'inspection', 'control', or 'safeguarding' have not been incorporated in any of the documents signed by Argentina and Brazil. 'Trustworthiness' and 'confidence' are, however, mentioned repeatedly. What can be derived from these documents is that *mutual trust* will be, or already is, the result of a series of activities, now underway, that those same documents mention: frequent visits, in-depth contact at a political and technological level, significant exchanges of information, industrial and technological complementation, reciprocal knowledge of their respective nuclear programmes, undertaking of joint projects of significant importance (such as that related to fast breeder reactors), and so forth, with everything co-ordinated and promoted by the Permanent Committee on Nuclear Policy.

It is the undeniable right of every country to determine by itself, without external pressure or influence, which means are adequate and sufficient to satisfy its own national requirements in matters of security and to have *trust* in the intentions and activities of another or other countries. Argentina and Brazil have embarked upon a process that includes, among its goals, addressing their legitimate security concerns. It is their right and responsibility, and only theirs, to determine which is the best, or at least the most suitable, manner for achieving them. On the other hand, we must not forget that every process is dynamic, not static, and that, as a result, the means of obtaining the sought-after outcome might evolve, because one means might prove inadequate or insufficient as better ones are developed or because existing ones have to be expanded or perfected. Still, in each case the decision on these matters lies exclusively with the interested parties.

The above discussion has to do with the view of those who see Argentine-Brazilian nuclear cooperation as a means of venting international suspicions of the ultimately non-peaceful intentions of the nuclear programmes of both countries, in other words, as a means of generating *international confidence*.[10] Undoubtedly, this confidence should be the logical result of the policy of cooperation. However, it is a mistake to believe that it is the main purpose of that policy, at least in matters of security. In this area, the goal is for Argentina and Brazil to protect their *individual* security and provide *reciprocal confidence*. International security should follow as a natural and desirable corollary, but it is not the determining factor, nor is it even publicly stated as one of the elements, among many others, that influences the *rapprochement*.

(4)   *International perception of the process.* From the beginning, the rest of the world, especially the developed countries, has watched the Argentine-Brazilian *rapprochement* on nuclear issues with interest. They considered the initial steps of the *rapprochement* to be promising and part of a process to promote mutual trust. At the same time they had doubts as to the continuity and real effectiveness of the process, especially on the part of Brazil, where they saw the military's influence on the nuclear programme as important.[11]

The international community's perception of this process is coloured by two factors. First, it has never fully understood that the agreements of nuclear cooperation occurred in the context of a much broader cooperation that encompasses many commercial, economic, industrial and other aspects. Second, its view of the nuclear cooperation between both countries is predominantly very narrow, seeming to interpret the main goal as being that of establishing a system of *mutual inspection*. A typical expression of this attitude is the following: 'There is speculation that the incoming Argentine Government may be less favorably disposed to international safeguards than its predecessor. This might mean that the plans for reciprocal inspection between Argentina and Brazil are not pursued.'[12]

I have already referred to these two issues in this paper. I can only add that the meaning of Argentine-Brazilian cooperation has been further distorted by a tendency by some to conceive of the supposed *reciprocal verification* within the parameters of the Nuclear Non-Proliferation Treaty (NPT), the Treaty of Tlatelolco and the system of safeguards of the International Atomic Energy Agency. In other words, the international perception of this bilateral process often embodies a belief that the end result should be a control regime similar or equivalent to that resulting from the above-mentioned documents.

Whatever the advantages or defects of those documents, which are not analysed here, we must not forget that the measures they contain are not the only possible ones – nor perhaps the most productive ones – for reaching the goals of providing confidence and security. Every case is not the same or similar. Procedures are convenient in some circumstances and not in others. The belief in solutions of universal application, based on models that work or could work well in certain cases, is in reality a simplistic approach that not everybody – each conditioned by his or her own political, technical and geographic particularities – can accept with equal ease.

It should not seem strange that some countries look for their own solutions to their own problems. Their answers may or may not coincide with those found in other cases, or may sometimes be adapted or modified totally or partially. What is important is that whatever the answer may be, it should provide confidence and security, that is, it should be satisfactory in the first place to the countries directly involved, which are the ones that must make the decisions, and then, as a consequence, have a positive effect on the rest of the international community.

In turn, the international community must follow the developments in Argentina and Brazil with interest and good will. No one can deny that the results being obtained by both countries are favourable. Two recent comments summarize this fact. First,

> The danger of nuclear weapon proliferation in Latin America has been dampened by an improvement of political relations between Brazil and Argentina. A regional policy centered on economic cooperation, particularly in the nuclear field, seems to be replacing the rivalry between the two countries, based on nationalistic military considerations.[13]

Second, and introducing a comparative element,

> Much less complicated than nuclear weapon free zones but still very useful are bilateral consultation and confidence-building visits of the sorts that have recently been undertaken by Argentina and Brazil.[14]

Faced with the interest with which this process is being followed and with the expectation with which spectacular advances are sometimes awaited, the process of Argentine-Brazilian nuclear cooperation continues to advance, slowly but effectively, with the construction of a series of ties that link the nuclear policies and achievements of both countries with increasing strength.[15]

What has already been achieved is very important and could not have been predicted only 10 years ago. Much is left to be achieved. The process is being pursued with energy and foresight. The goals of peace, security and development are indisputable. Their achievement will benefit not only

Argentina and Brazil, but also the other Latin American nations and, indeed, the whole world.

Response – *Dr Oliveiros S. Ferreira*

Etiam diabolus audiatur
(Let even the devil be heard)

One has only to focus on the total meaning of the epigraph with which I head my comments on the brilliant presentation of His Excellency Ambassador Julio Carasales to take from this seminar the conclusions that prevail.

Ambassador Carasales took care to begin his lecture by saying that he speaks for himself and that even the added emphasis placed on certain words or sentences were exclusively his responsibility. These precautions did not prevent him from expressing certain truths that should be noted for a perfect understanding of the issues covered by the general subject of the seminar, 'Latin American Nuclear Cooperation: New Prospects and Challenges'. I have one advantage and one disadvantage in relation to Ambassador Carasales. The advantage is that, not belonging to any institution that may take offence at what I say here, I believe I have the freedom to speak what might be considered as improprieties. The disadvantage is that, by speaking improprieties, I run the risk of being taken for what I am not. Every time I speak to a Latin American group, especially Uruguayans and Argentineans (and fellow citizens from the great fatherland that Simon Bolivar and Artigas dreamt of), I feel obliged to make it clear that I am not, and never have been, what many years ago the journal *Marcha* called me and that was later repeated in Buenos Aires and other capitals. I simply want to say that I am not a spokesman for the armed forces of Brazil. I speak here as a professor, journalist and person interested in military matters. Having made that clarification, let me proceed to the facts.

Before directly commenting on Ambassador Carasales' presentation, allow me to say that this seminar proves what is said in textbooks on international relations – that international relations between two countries actually involve at least three if not more than four countries. We are here now in Montevideo at a seminar sponsored by the Nuclear Control Institute of Washington, D.C., USA to discuss, especially today, 'What the objectives of Brazilian-Argentinean nuclear cooperation are, and what political *rapprochement* must take place for these to be achieved?'

THE US FACTOR

With all due respect to my noble colleague who preceded me, I believe that we should, before anything else, look at the United States' perception of Brazil and Argentina's *rapprochement* in the nuclear field, and not just at Washington's perception of the problem, but also at the perception that Buenos Aires and Brasilia may have of the United States' interest in the issue and how they perceive the North American diplomatic movement.

It seems strange to begin my remarks with such a long aside. The reason is that what truly brings us together in this meeting is more than a desire to see Montevideo again or to get to know one another and enjoy the pleasures of reinvigorating intellectual contact.

I cannot go into the tortuous history of relations between Washington and Buenos Aires and between Washington and Brasilia (previously Rio de Janeiro) in detail here. I can, however, summarize it in one sentence: as far as it was able to, the US government attempted to *hold the balance* in the complicated relationship between Argentina and Brazil – a relationship in which Uruguay, we should recall, has always been a partner whether it wanted to be or not, since what we in Brazil call the Cisplatina War, fought between the Argentinean Confederation and the Empire of Brazil in 1825–6. The relationship between Brazil and Argentina has been adversarial ever since the Spanish and Portuguese crowns set foot on the banks of the River Plate. It has always been marked, as Ambassador Carasales said, by 'mistrust, competition and antagonism'. I therefore believe the political decision of President Carlos Saul Menem to repatriate the mortal remains of Don Juan Manuel Rosas will make nationalist hearts beat with greater vigour on both sides of the River Plate, in memory of the 1852 campaign.

Given this history, in which words of Latin American fraternity always came after aggressive acts of nationalist hostility, we should reciprocally acknowledge that the United States was always, for many sectors of what I call the cultured school of Brazilians, the country to be imitated, as well as the power that in the 1830s and 1840s was able to prevent the nationalism that Rosas, Facundo and later Peronism dreamt of, from imposing its hegemony on the River Plate basin. On the other hand, I believe I am not exaggerating if I say that there were moments during this century in which a *rapprochement* with the United States meant, for Argentinean leaders, the possibility of overcoming Brazil's hegemonic aspirations over all of Latin America.

Forgive me, Ambassador Carasales, if I touch on matters that diplomacy recommends be kept at the back of the drawer, including academic drawers. I do it with the certainty that I am committing one of the improprieties to which I referred at the beginning. If we do not exorcise the US ghost, we will not understand each other as nations. The best way to distance this ghost, which has brought us together here to tell it how we intend to co-operate in the nuclear area, is to recognize that it exists, that it intrudes in our affairs and that it considers us inferior. Perhaps the word 'inferior' is too strong. For the sake of fidelity, it would be better to replace it with another word that I heard a US diplomat in my country use: irresponsible.

One of the problems that always assails those who believe in power politics, as I do, is the development of international relations, or, rather, the increasing awareness that international law exists, despite those who like *Machtpolitik*. One of the problems for adherents of *power politics* is the certainty that the law imposes forced companionship between the 'responsible' and the 'irresponsible'. Easy were the days when Drago could protest against the British and the Germans who were bombarding Venezuela to collect debts, when the Baron of Rio Branco could refuse to discuss this minor question, and when Theodore Roosevelt could quietly say 'I took the canal.' Today, one does not make international policy with the same ease as was the case at the beginning of the century, especially because, to use Karl Deutch's expression, the power differential between the responsible and the irresponsible is getting smaller.

THE OBJECTIVES AND STEPS IN NUCLEAR COOPERATION

In my position as Ambassador Carasales' conference discussant, I should not move too far away from what he said, nor from the general subject of the seminar, nor from that of our work session. Besides, I believe that when I am asked a question, good manners obliges me to answer it.

The best answer to the question, 'What objectives does Brazil-Argentina nuclear cooperation pursue', Ambassador Carasales gave in his presentation when he quoted two Argentinean-Brazilian presidential statements – 'coordination of political positions in the international sphere for the defence of common interests' (the statement from Brasilia) and 'complete agreement on Argentinean-Brazilian positions in the nuclear area' (the statement from Ipero). I even dare to say that were it not for the insistence of the US government that Buenos Aires and Brasilia sign the Nuclear Non-Proliferation Treaty (NPT), both countries would not have felt the need to co-ordinate policies in the face of pressures that affect them equally.

The answer to the other part of the question is more complex – what are the political steps that must be taken to attain cooperation between Buenos Aires and Brasilia with regard to agreement on a common policy toward outside pressures to sign the NPT? The answer will vary, at least in Portuguese, depending on the meaning we place on the verb 'attain' (*atingir*). Does it mean to 'reach' or to 'hit'? I assume that in English this play on words does not pertain. I apologise, therefore, to the translators and to those who speak English.

Whatever meaning one may attribute to the word 'attain', it is the total Brazilian-Argentinean relationship that is at stake in the relationship in the nuclear field. Ambassador Carasales is fully aware of this fact, and the first part of his magnificent presentation was a diplomatic attempt to draw our attention to the delicate terrain that Presidents Raul Alfonsín and José Sarney stepped onto and that Presidents Carlos Saul Menem and José Sarney are now walking. I draw your attention to two passages from his presentation that perhaps went by unnoticed. I begin this sequence of quotations with an observation on the style that fills Parkinson's tasteful books on the laws governing bureaucracy. It is interesting that Ambassador Carasales should have quoted, in support of the need for cooperation, a speech by the Brazilian Ambassador to Argentina, Antonio Azeredo da Silveira, who, after becoming Brazil's foreign minister, put into practice the policy of 'big power chauvinism' developed by the Geisel government against Argentina with regard to the Corpus-Itaipu question.

However, let me go on to quote certain passages. 'Nobody can deny the fact that for many years the military in both countries always considered as a priority in their hypotheses of conflict the possibility of a conflict with the neighbouring country.' Other quotes are: 'If this process [of integration] is definitively consolidated and a common market is created'; and 'past experience also weighs heavily in an evaluation of this Argentinean-Brazilian process: it is characterized by more than a century and a half of conflict and rivalry and by the failure of some efforts at *rapprochement* in the face of the disappearance of the political figures that promoted them'. These references are enough to indicate that one cannot suddenly wipe out from the history of two peoples a century and a half of suspicions, above all when in most cases the core of the armed forces, if not the foreign ministries, harbour these suspicions.

Possibly tourism has today eliminated many points of friction between the two peoples. I remember, however, that in 1973 in Buenos Aires I was embarrassed by the big power, chauvinistic behaviour of some Brazilians, and during the same period I was irritated to see that some Argentinean magazines called Brazilians 'little monkeys'.

The truth is – and Ambassador Carasales takes this point into account – Brazil and Argentina are playing all the cards of nuclear cooperation in the eventuality of success of cooperation in the economic area. One lesson I learned as a student was from Emile Durkheim, who discussed the question of the influence of transportation (material density) on the cohesiveness of a population. The French master pointed out that in England railways served to transport merchandise and facilitate business, but not to bring people together, the opposite of what happened in France.

It is too early to say what the result of economic integration between Brazil and Argentina might be. Will it lead to a greater *rapprochement* of peoples and to an overcoming of the national antagonism that some Argentinean nationalists and Brazilian liberals defend? On the other hand, will these forces lead only to protests by the planters of wheat and apples and the producers of cheese and wine in Brazil against the Argentineans, who will see the efforts of Brazilian civil engineering firms to penetrate the Argentinean market as a motive for renewed protests and more intense hostility. As long as the problem is only about cheese and wine, then with pleasure I cede primacy to Argentina. It will be easy to resolve our problems. The whole question – and the reason we must give the devil the right to be heard – is in the hands of the armed forces. The nuclear programme is completely in their hands. If anyone doubts that fact, it is surely not the concerned Washington diplomats.

PERCEPTIONS OF FOREIGN POLICY

The history of relations between Brazil and Argentina amounts to an account of the perceptions that the national groups most able to voice their conceptions of foreign policy have of one another. What I call 'critical realism' in the evaluation of relations between the two countries often leads me to support positions that are in conflict with those expressed by Ambassador Carasales. I apologise in advance for bringing to the debate diverging visions of the history of both countries. However, the two groups will fundamentally agree, I believe, on the problem of nuclear cooperation.

Who in Brazil voices a conception of foreign policy? The nationalists and those I call liberals do, taking into consideration that in Latin

America liberals are basically illuminati. To the former, the nationalists, the *rapprochement* with Argentina is important because it will allow the construction of a bloc in opposition to the United States. For the latter, the liberals, Argentinean nationalism is the danger to be confronted – the nationalism that comes from Rosas and continues with Peron. For this reason, the liberals see the United States as far from being a strategic adversary. Rather, it is a great tactical ally that is sometimes also a strategic ally. Thus it is understandable that the defenders of the *rapprochement* between Argentina and Brazil may have been the presidents who in one way or another became affiliated with the Brazilian nationalist movement – Julio Vargas, Juscelino Kubischek and Janio Quadros, who sought in Argentinean antagonism against the United States common points for supporting a foreign policy that was intended to be independent.

It should be emphasized that history moved the Brazilian liberals away from the forefront and took from them the possibility of assuming power and carrying out their foreign policy. This fact must be taken into account, in association with another fact that Ambassador Carasales emphasized in his talk and to which I return from another perspective: the position of the armed forces.

From Vargas to Quadros, the doctrine of the Brazilian armed forces has continued to be that inherited from the Empire, even though in 1931 General Pedro Aurelio de Goes Monteiro assigned to the army a new mission, in view of the declared impossibility of conflict at the borders: to occupy the vast territory of Brazil. Is it not strange, therefore, that the post-1964 governments up to that of President Sarney brought relations with Argentina to their highest point of conflict? It should be emphasized, for the sake of truth, that reaching this point, almost of no return, contributed greatly to Argentinean military nationalism, which, in its geopolitical watershed, found its greatest proponents in General Gugliamelli and President General Lanusse, who knew how to push to the limits of diplomatic aggravation and to stop at that point, having made their hostile intention known.

Given that past, there is nothing to indicate that the armed forces of either country have abandoned the prejudices inherited from the Empire and the Argentinean Confederation. On the contrary, I would even say that the sincerity of Presidents Alfonsín and Sarney in attempting *rapprochement* between the chief of staffs by carrying out 'Symposia of Argentinean-Brazilian Strategic Studies', sponsored by the Joint Staff of the Armed Forces of Argentina and by the staff of the Armed Forces of Brazil, has tended to fall by the wayside as a result of a lack of funds, the earnestness of the presidencies themselves, and a lack of vision, or concern over not having vision, of the individual armed forces staff, who are the ones who decide.

Economic integration does not respond to the evolution of ideas and military doctrines in the armed forces – a point I think I can make about both countries. The efforts to carry out economic integration in the Plate region in my opinion respond more to the sincere desire of the heads of state to find a way out of the crisis of the western world, especially Latin America, brought on by the oil situation and related interest payments than to the need for growth in both economies or for a project of common destiny forged by the everyday affairs of the border populations and the principal promoters of unity of both nations. In Buenos Aires, at the first of the symposia to which I referred, I supported the idea that sovereign nations can only consider military cooperation after their economies are in fact integrated, a state that would mean that the suspicions regarding the partner's motives had disappeared, inasmuch as they are economically based.

If this hypothesis is correct, it is still too early to bring cooperation in the nuclear sector to the negotiating table, since this area is a most sensitive one for the respective military establishments. If the purpose is to give the impression to the world and to the armed forces themselves that there is movement toward overcoming the almost two-century-old differences, it is strange that the proponents are trying first to achieve the absolute rather than proceeding through ways that are less sensitive. With respect to difficulties in the economic area, I quote just one sentence from Ambassador Carasales: 'Moreover, it is wrong to ignore existing problems, especially in the economic and commercial spheres; while not addressed in this paper, overcoming them will require energetic and imaginative measures.'

Referring to the military sector, I would like to analyse some of Ambassador Carasales' observations, which I believe are meaningful to those who understand their intent.

It is not my purpose to insist on the peaceful nature of the Argentinean and Brazilian nuclear programmes. Those who follow events closely, or even from afar, consider this type of declaration to be merely rhetorical and without political consequence. Possibly because of this attitude, President Mikhail Gorbachev has been more prudent than other heads of state, making it clear in the Moscow Declaration, signed by him and President Sarney when the Brazilian president visited Moscow, that the two heads of state agreed to exert the greatest efforts to avoid nuclear proliferation. For a government such as Brazil's, which insists on not signing the NPT, this declaration is at the very least curious, even though it may be a response to those who fear proliferation in irresponsible hands.

The point I want to emphasize in Ambassador Carasales' lecture is, however, something else. It should be noted that his words evince the expression of a new doctrine. It is very subtle but enhances an understanding of the objectives of Argentine-Brazilian cooperation, apart from the objective I referred to earlier of agreeing on a common foreign policy on the subject. I conclude from the words of Ambassador Carasales that the documents signed since 17 May 1980 have the objective, among other things, of strengthening reciprocal trust between the Argentinean and Brazilian governments, and, to speak another impropriety, I would go as far as to say between the armed forces of the two countries. Another constant element of these agreements to which Ambassador Carasales draws attention is the concern with peace and security in the region. The hypothesis that the armed forces do not adhere to this policy is too serious and for this reason should be rejected.

I would say that the Argentinean armed forces staff is aware of Brazil's naval inferiority, while the Brazilian armed forces staff is aware not only of the superiority of Argentina's Navy, but also that its armament (maintained at the level before the 1982 war) cannot confront a Soviet threat in the South Atlantic, which would involve a submarine war. I would also say that the intelligence services of the two fleets monitor with special interest the progress each country makes in the nuclear area and knows exactly the point reached by each research laboratory. The Brazilians also know that Argentinean progress, achieved outside the control of the International Atomic Energy Agency, is superior to Brazil's.

The contours of this situation are determined by the following factors: (1) a desire for economic integration, which can be speedily undertaken; (2) ignorance of the possibilities of cooperation in the conventional military terrain (where the hypotheses of conflict were always formulated); and (3) a quest for cooperation in the nuclear area. In this scenario, there is insistence on the need for nuclear cooperation to reinforce mutual trust and guarantee peace in the region.

It is easy to verify how difficult it is to attain compatibility among these factors. Yet it can be attained if a key word is introduced in the decodifier: deterrence. Only deterrence as a military doctrine leads to an understanding that if you wish to know what the other has, you in turn inform him of what you have. Deterrence, however contradictory the statement may appear to be, does not lead to war but is one of the guarantees of peace. Let the European peoples speak, who up to now have lived in peace thanks to the existence of ballistic missiles and the bomb.

This position is clear in Ambassador Carasales' presentation. Regarding security, his Excellency said, 'the objective is to protect one's own security, by providing *reciprocal trust*'. Because each party wants peace but always suspects the other's intentions, they decide to exchange information and form a common front against those who wish to prevent each one from developing an autonomous provision for security. I refer to one of the final points in Ambassador Carasales' presentation: 'Argentina and Brazil have embarked upon a process that includes, among its goals, addressing their legitimate security concerns. It is their right and responsibility, and only theirs, to determine which is the best, or at least the most suitable, manner for achieving them.'

Yet it must be observed that this declaration of sovereignty, in a world where remote sensor capacity satellites allow a country to get to know its neighbour's soul, injures third parties, as is inevitable. While the doctrine of deterrence can be compatible with the objective of giving reciprocal trust to Brazil and Argentina, it may be also a central element in a strategy for all the azimuths, as General Charles de Gaulle would say, or any other adversarial strategy with real interests or *suppositions* of hegemonic power in the hemisphere. Possibly Washington does not believe that Brazilians and Argentineans decided not to make war. Possibly Washington is afraid that a conflict could take place involving the use of an atomic weapon, a step that would destabilize the US rearguard at an internationally delicate moment, given the instability in the USSR. Possibly there is a fear that less responsible people will use the nuclear submarines being studied in this strategy for all the azimuths. Perhaps because of all of these fears, and so as not to lose its role as *holder of the balance* between Brazil and Argentina, Washington is so very interested in Brazilian-Argentinean cooperation and is pressuring the Brazilian government to sign the NPT.

I think I have spoken too many improprieties, even though they are pertinent to a discussion of the subjects proposed to us.

With the clarity and objectivity that marked his presentation, Ambassador Carasales, diplomatically, said much more than I did. You will have understood, I hope, that he and I agree on one point: we know that, whether they are irresponsible or not, the governments of Argentina and Brazil want to find their own answer to their own problem. We also know that the inspection parameters set by the NPT, by the IAEA and by the Treaty of Tlatelolco are not the only possibilities and that it is up to the countries interested in re-establishing trust among themselves to find an answer to their own problems.

I appropriate the words of Ambassador Carasales to end my presentation: ' . . . whatever the answer may be, it should provide confidence and security; that is, it should be satisfactory in the first place to the countries directly involved, which are the ones that must make the decision . . . '.

Discussion – *Dr John R. Redick*

I have listened to the presentation of Ambassador Julio Carasales and to the remarks of Dr Oliveiros Ferreira with great interest. I want to focus my remarks on Ambassador Carasales' paper because I had an opportunity to read it in advance. I compliment him on a thoughtful and provocative paper. I agree with both its tone and its content. What I have to say will build on some of the points he made.

I strongly agree with an important theme running through Ambassador Carasales' paper – Argentine and Brazilian nuclear cooperation must be understood in the context of the economic framework emerging between the two nations. Nuclear cooperation between Argentina and Brazil is not 'the tail wagging the dog'. It is part of a broader web of relationships that can benefit both nations.

In his paper, Ambassador Carasales correctly compares the Argentine-Brazilian relationship to the *rapprochement* between the Federal Republic of Germany and France that occurred after World War II and extended into other political areas. This *rapprochement* paved the way for what we are now witnessing, a developing European Economic Community.

I suggest that this European analogy could be carried into the nuclear sphere as well, recalling that the European *rapprochement* also led to the creation of the European Atomic Energy Community, or EURATOM.

I suggest that this model is an appropriate one for Argentina and Brazil and all Latin American nations to consider. I am aware that Argentina and Brazil have suggested that their bilateral nuclear relationship, as it evolves, would develop outside of the International Atomic Energy Agency (IAEA) safeguards. I submit, however, that the EURATOM model, by which European nations concluded a single agreement with the International Atomic Energy Agency for the entire region, may have relevance. I hope this approach will be a topic of discussion at this conference and elsewhere.

My second point focuses on Ambassador Carasales' comments regarding the evolution of the Argentine-Brazilian nuclear relationship into a bilateral system of mutual inspection and verification. He correctly noted that none of the agreements signed since 1980 mentions the word 'inspection'. I also agree that a great deal of attention – perhaps excessive attention – has been given outside of Latin America to the idea of inspections, particularly by countries that have nuclear weapons and that have not always done a very good job of controlling or reducing them. Certainly, the external speculation on the point of inspections has exceeded the present reality.

I do, however, question the implication that inspections or verifications are something that were invented elsewhere and superimposed on the bilateral discussions between Argentina and Brazil. My understanding is that Argentine and Brazilian officials have discussed joint research and information exchanges on monitoring equipment and reactor fuel burn-up rates. Now, this process is not an inspection per se, but it is the type of cooperation that is fundamental to a safeguards regime – that is, non-intrusive monitoring systems, state systems of accounting. As we all know, these are part of the IAEA regime, at least as much a part of that regime as are physical inspections.

I concur with Ambassador Carasales' view that we must not interpret the main goal of Argentine and Brazilian nuclear cooperation as the establishment of a mutual inspection regime. However, I also believe that some sort of regularized or systematized process would deepen and reinforce the broader objectives of Argentine-Brazilian cooperation and help it to continue, even in the face of the political shifts and changes that occur in all countries having democratic and pluralistic systems.

My third point concerns the relationship of Argentine-Brazilian nuclear cooperation to the non-proliferation regime. As regards the Nuclear Non-Proliferation Treaty (NPT), which Argentina and Brazil do not accept, I would only point out that it is entering a very important time. We are moving toward a 1990 Review Conference and, in 1995, an important extension conference that will determine the future of the NPT. Between now and 1995, there will be critical debates and, we hope, major progress by the nuclear weapons states to reduce their nuclear weapons and, indeed, reach a comprehensive test ban treaty.   I would urge that Argentina and Brazil use what Ambassador Carasales refers to in his paper as the 'coordination of common foreign policies in nuclear matters' to do more than simply criticize the NPT, but also to suggest creative and constructive changes. For example, last year the Brazilian foreign minister, in a speech to the Geneva Conference on Disarmament, called for 'restoring a consensus on non-proliferation as it was originally established in favor of a more equitable and less oligarchic model than that established in the NPT'. This statement is interesting and we can rightfully ask for more elaboration – for new ideas and approaches.

As the Argentine-Brazilian *rapprochement* deepens and nuclear cooperation becomes even more serious in the future, I believe those countries' criticisms of the NPT and their ideas as to how it might be changed will be listened to with great seriousness.

With regard to the Tlatelolco Treaty, because a part of Panel 6 is devoted to it, I will make only one brief comment. I remind the participants that Argentina and Brazil took a major role in framing the original agreement and that Tlatelolco has certain remarkable features that have not been developed. Currently efforts are underway within the Organization for the Prohibition of Nuclear Weapons in Latin America (OPANAL) to make Tlatelolco more attractive to Argentina and Brazil in the hopes that those nations will become full parties. Argentina, in particular, has raised a number of issues that should be discussed as it considers ratification. It is time to revisit Tlatelolco with an open mind for compromise. Anything is possible, including some revision of the agreement, so long as the fundamental core of Tlatelolco is preserved.

I close with one modest suggestion. I refer to one part of the nonproliferation regime that is not frequently discussed, and that is the Limited Test Ban Treaty of 1963. Argentina and Brazil are parties to the Limited Test Ban Treaty. Both countries have also consistently recommended that the superpowers develop a comprehensive test ban treaty to prevent explosions of all nuclear weapons underground. I agree with this recommendation. Last year, the Brazilian foreign minister called for the creation of a special UN committee to negotiate a comprehensive test ban agreement.

As many of you know, there is now an effort to revise the Limited Test Ban Treaty, to amend it and make it a comprehensive test ban agreement. There will be an international conference early next year to consider this suggestion. The nuclear weapon states are resisting this effort – unwisely, in my opinion. My modest suggestion is that Argentina and Brazil consider a bilateral diplomatic initiative to extend *unilaterally* the Limited Test Ban Treaty limits into a fully comprehensive test ban treaty between them, and then to challenge the United States and the Soviet Union to follow suit.

In doing so Argentina and Brazil would be providing a constructive model for the nuclear weapons states. Moreover, it would emphasize, to the nuclear weapons states, just how vitally important the leading non-nuclear weapons states view the halting of all nuclear testing. It would deepen the ongoing bilateral nuclear cooperation between Argentina and Brazil. The two nations would give up little because, as the nuclear weapons states have by now learned, peaceful nuclear explosions are a misnomer.

They are not environmentally safe, and they serve no practical peaceful purpose.

Discussion – *Ambassador Hector Gros Espiell*

It is a pleasure for me to be here with you today to discuss a subject that, for many years, was at the centre of my own concerns. It is a subject that continues to be of very special interest to me. The approach I want to take is one that derives from my focus on the issue of nuclear cooperation for Latin America and for Uruguay in particular.

In my view, an analysis of this issue leads to a series of inescapable conclusions. First, Latin American nuclear cooperation is an essential and indispensable element for development, stability and peace in the region. However, nuclear cooperation can only take place within a framework of general cooperation, and specifically economic cooperation, between Argentina and Brazil. In turn, cooperation, understanding and confidence between Argentina and Brazil are, I think, keys to the future of Latin America.

As a Uruguayan, I believe that Uruguay, because of its history and political relationships, has a unique role to play in the evolution of relations between Argentina and Brazil. Uruguay came into independent existence as a result of a war between Argentina and Brazil. The 'birth certificate' of the Uruguayan nation is a peace treaty between Argentina and Brazil, the preliminary peace treaty [Convencion Preliminar de Paz] of 1828. The only international war in which Uruguay participated actively was the Triple Alliance War, in which Uruguay acted as an ally of Argentina and Brazil against Paraguay. From the time Uruguay became an independent state until 1870, its politics were intertwined with Argentine and Brazilian politics. For Uruguay, the question of economic and nuclear cooperation between Argentina and Brazil is very much a Uruguayan problem. We see it not from the outside but very much from the inside.

Ambassador Carasales pointed out, in my view quite accurately, that nuclear cooperation between Argentina and Brazil cannot be studied or analysed outside the framework of overall cooperation between Argentina and Brazil. Nuclear cooperation is a part of this broader cooperation. It can only be understood, it only has significance, within this broader context, within this framework and within an understanding of the general history of the relationship between Argentina and Brazil. Nuclear cooperation is therefore an aspect of the more general cooperation that has been taking place in recent years. It can only be understood from the perspective of the process of integration and economic cooperation that marks the most recent relations between the two nations.

The ongoing economic integration of Argentina and Brazil is essential to understanding the problems of nuclear integration. Here the Uruguayan perspective is unique because Uruguay has, in fact, been participating in the economic integration, which is no longer a bilateral process but rather a trilateral one. Moreover, the economic integration is open to other Latin American countries. Uruguay, as a partner in the economic integration, would ultimately be involved in nuclear cooperation.

An initial comment I would like to make with respect to nuclear cooperation between Argentina and Brazil is to affirm that it is geared to specific peaceful purposes. This conclusion is affirmed by all the legal instruments executed over the course of this process in recent years. While these are not treaties from an international point of view but rather joint declarations, we cannot ignore their international legal value, as they constitute the two countries' obligations to each other. On the question of nuclear testing in Argentina and New Zealand, the International Court of Justice ascribed to the unilateral declarations made by France the force of international law, by virtue of their having been official statements of the President of the French Republic. Thus, it seems to me that this factor is very important because these statements, these declarations, generate international obligations. It is true that Argentina and Brazil are signatories but not parties to the Tlatelolco Agreement (although Brazil has ratified it). However, Article 18 of the Vienna Convention on the Law of Treaties establishes that a signatory country, even if it is not a party to a treaty, is obligated not to do anything contrary to the objective and purposes of that treaty.

I would like to conclude by saying, first, that this process of nuclear cooperation is essential for Uruguay and for Latin America as a whole. Second, I have no doubt as to the peaceful purposes of this process. Finally, we must find a way to make this process subject to a system of international control. The system does not necessarily have to be that of Tlatelolco or any other existing treaty. However, we must ensure that the process is part of an enormous effort for peace.

3 Industrial and Economic Benefits of Latin American Nuclear Cooperation

Presentation – *Fernando A. S. Henning*

NUCLEAR TECHNOLOGY IN BRAZIL

Government Policy on Nuclear Energy in Brazil

Constitutional aspects

Before I address nuclear cooperation in Latin America, it is important to understand the Brazilian government's policy on nuclear energy. It is based on the following principles, derived from Brazil's constitution:

- To ensure the country's sovereignty.

- To further the fundamental aims of national development, eradication of poverty and reduction of social and regional discrepancies.

- To support the principles underlying Brazil's international relations – national independence, self-determination of nations, non-intervention, equality among nations, defence of the peace, peaceful resolution of conflict and cooperation among nations for the progress of mankind.

Two other points are important. First, the state is responsible for inspecting nuclear services and installations of any nature, for exercising a monopoly over research, mining, enrichment, reprocessing, industrialization and commerce in nuclear minerals and their by-products, within the context that:

- all nuclear activity in Brazil should be for peaceful uses only, subject to approval by the National Congress; and

- the state will authorize the application of radio-isotopes for research and use in medicine, agriculture, industry and analogous activities.

76

Second, all people have the right to an ecologically balanced environment, which is for the common use of the people and is essential to a healthy quality of life. It is the responsibility of the government and the community to defend and preserve that environment for present and future generations.

Basic guidelines
Within the above framework, Brazil's policy embodies the following aspects. First is the *sovereign right of use*: Brazil has the right to use nuclear energy as one of the indispensable tools for its development. Second is *respect for nuclear agreements*: Brazil will honour all international agreements entered into in the field of nuclear energy, especially those related to the safeguards of the International Atomic Energy Agency (IAEA). Third is the *non-proliferation of nuclear weapons*: Brazil will defend the principle of non-proliferation of nuclear weapons, horizontally as well as vertically, and always on a non-discriminatory basis. Fourth is *international cooperation*: Brazil will promote cooperation with the international community to improve national technologies and industry and to intensify scientific, technological and industrial exchanges. Fifth is *protection of national science, technology and industry*: the government will protect the interests of Brazil and its citizens with respect to any advances in the nuclear field. Sixth is *exporting for peaceful purposes*: through international agreements Brazil will guarantee that countries to which it exports, directly or indirectly, raw materials, technology, knowledge and equipment will not use them to produce nuclear weapons.

Objectives of Brazil's nuclear energy policy
A number of objectives underlie the evolution of nuclear activities in Brazil:

- Development of a complete fuel cycle at the scientific-technological and industrial levels.

- Establishment of a domestic scientific-technological and industrial capability in the production of materials and equipment for nuclear energy.

- Development and implementation of projects involving the nuclear reactors, power plants and installations necessary to carry out the programmes deriving from Brazil's nuclear policy.

- Promotion of the use of nuclear technology for Brazil's economic and social development, especially in the areas of health, agriculture, industry, energy and environmental protection.

* Transfer of technology to the industrial sector

* Priority for the research, mining, industrialization and storage of nuclear minerals, as well as other minerals necessary for nuclear energy, as required to meet the country's demand, and export of any surpluses.

* Safe operation of nuclear and radioactive installations, as well as of activities involving ionizing radiation.

* Development of the scientific-technological and industrial foundation needed for nuclear energy, with support for research and development institutions and for scientific-technological and industrial exchanges with other countries.

* Adequate build-up of human resources, compatible with Brazil's needs in the nuclear field.

* Provision of accurate information about the benefits, risks and safety measures associated with the use of nuclear energy nationally.

* Adequate disclosure of activities involving nuclear energy occurring in Brazil.

The national Congress is now analysing Brazil's nuclear policy. Once approved, it will become official policy and implemented as such.

Overview of Nuclear Energy in Brazil

The academic and theoretical studies on nuclear energy being carried out in Brazil assumed political and economic significance in the early 1950s when the government decided to control the export of nuclear minerals, whose production was not subject to regulation. At the same time, there was growing interest within Brazil in promoting a national nuclear technology. To enhance the standing of nuclear research, in 1956 the National Commission of Nuclear Energy (CNEN) was placed under the Presidency of the Republic.

Since the early days, nuclear development in Brazil has involved the pursuit of both external cooperation and an indigenous programme. Brazil's first research reactor was purchased from the United States; it began operating in September 1956 at the Institute for Energy and Nuclear Research (IPEN) in São Paulo. In 1960 a second research reactor started operations at the Institute of Radioactive Research, now called the Centre for Development of Nuclear Technology (CTDN), in Belo Horizonte. In 1965 Brazil, with support from the United States, completed construction of the Argonaut, the first Brazilian research reactor to have a considerable proportion of national components; it is located at the Institute of Nuclear Energy (IEN) in Rio de Janeiro. That same year Brazil signed an agreement for nuclear cooperation with the United States.

In 1968 Brazil decided to include the nuclear option in its energy future. To this end, it planned to construct and operate its first nuclear power reactor in order to gain experience in their construction and operation. In 1972, under an extension of an existing agreement with the United States, the Brazilian government awarded a contract to Westinghouse to build the Angra 1 nuclear power plant.

The low level of participation by domestic industry in Angra 1 led the government to pursue a programme of development of nuclear power plants that would provide a market for industry in this area in the long term. It realized that its objectives of technology transfer, of implementing a complete fuel cycle and of promoting greater participation by domestic engineering firms and the nuclear industry would not be met by relying on international bids for the construction of power plants, an approach that would result in an arbitrary schedule of construction and in plants with different technologies. Moreover, although the technology for the design, construction and management of conventional power was readily available, it was not transferable to nuclear plants, whose operational safety would be jeopardized.

In keeping with its nuclear policy, the government adopted a strategy that entailed a programme under which the introduction of nuclear power plants for the production of electricity would be started parallel with the development of fuel cycle technology so as to achieve independence from abroad as soon as possible. To identify the parameters of the programme, the government undertook a feasibility study, taking into account the following:

- inclusion of the nuclear programme under national planning for the generation of electricity;

- coordination of the programme with implementation of the fuel cycle;

- coordination of the programme with participation by the national engineering and nuclear industry;

- identification of the required human resources and their education and training;

- technology transfer; and

- financial resources.

The results of the study led to the choice of a pressurized light water reactor (PWR) with a standard size of 1300 MWe. This decision required that the government negotiate a comprehensive nuclear energy programme, with foreign participation resulting in the transfer of technology for power plants and the fuel cycle.

Brazil's Nuclear Programme

The agreement with the Federal Republic of Germany (FRG)
The FRG, which had been co-operating with Brazil in the nuclear field since 1969 under a general agreement on cooperation in the areas of scientific research and technological development, had the capability to meet Brazil's desire to obtain the technology essential for implementing an autonomous nuclear industry, including the fuel cycle. In June 1975 the two countries signed an agreement on the peaceful use of nuclear energy, and later that year the national Congress approved it. At the same time, the Ministry of Mines and Energy of Brazil and the Research and Technology Ministry of the FRG signed an industrial protocol specifying the guidelines for each area of cooperation. These intergovernmental agreements were complemented by conventions between a state-owned Brazilian company, NUCLEBRAS, and several foreign companies to set up joint ventures that would result in transfers of technology and equipment in each area of cooperation.

The nuclear power plant programme under the agreement
Key provisions of the programme were as follows. With respect to the construction of eight standard 1300 MWe PWRs, Brazil would begin by supplying 30 per cent of the electromechanical components for the first two plants and finish by supplying 70 per cent of those components for the last two. An integrated engineering company, NUCLEN, a joint venture of NUCLEBRAS and SIEMENS/KWU, would be created to design and manage the construction of the nuclear power plants, while another firm, NUCLEP, would be set up to produce the nuclear steam supply system (NSSS) heavy components. In the case of the first four plants, all imported components would come from the FRG; in the case of the last four, Brazil would ask for international bids. Brazil also planned to determine the level of its domestic uranium reserves. Finally, Brazil would develop a domestic

industrial capability for the entire fuel cycle, from the production of yellow cake to enrichment and the fabrication of fuel elements.

Development of the Industry under the Programme

Establishing parameters

In defining a strategy for enhancing the participation of Brazilian industry in the programme, Brazil looked at several factors. First was the market – what materials, equipment and the like would the nuclear power plant programme require. Given the detailed plan for constructing the plants and assuming its timely implementation, it was possible to identify and quantify the expected demand for materials and equipment. Second was supply – what was the existing capacity of Brazilian industry and what was the potential for promoting new suppliers (this aspect of the study lasted a year)? Third, what were the critical constraints in terms of economics, technical and industrial technology? Fourth was the economics. The main issues were identified to be the size of the market and the need for investment funds to ensure local participation and to pay for the equipment to be supplied. The government needed to take measures to promote continuous participation by domestic firms in the market and to support the investments required both to purchase foreign technical assistance and to effect changes in existing equipment and procedures. The expected technical problems related to the low capability of the local engineering industry, the need to improve planning, production and controls, and, most of all, the absence of quality assurance procedures. A further point was the difficulty of adapting the ASME codes prevailing in Brazil's industry to the FRG's codes (DIN). With respect to the technologies to be developed, the specific needs were to strengthen Brazil's welding technology, to upgrade its capacity for non-destructive testing and to train the required labour force.

Strategy for promoting the participation by and development of industry

The government used nine key initiatives to raise the level of participation of Brazilian industry in the nuclear power plant programme. (1) Over time and through technology, it fostered the development of a domestic equipment industry, based on the construction programme with its target of 30 per cent Brazilian components for the first two plants. (2) NUCLEBRAS initiated a systematic programme of registering and qualifying industries; NUCLEN later took over this programme. (3) CNEN, the licensing authority, had already adopted the thirteen safety criteria of the IAEA and licensing procedures similar to those of the US Atomic Energy Commission. However, given the use of West German technology for the PWR programme, the government decided to establish an independent inspector. To this end, it set up the Brazilian Institute for Nuclear Quality (IBON), with representatives from government, industry and technological organizations. (4) Because Brazil's nuclear industry did not in general have quality assurance procedures, it was necessary to establish a quality assurance system that the architect engineer (NUCLEN) and the independent inspector (IBON) were to follow. (5) NUCLEN, which had responsibility for technology transfer and industrial promotion, assisted industry in: interpreting technical specifications and adapting the procedures for equipment design and production to nuclear norms, especially regarding safety; elaborating quality assurance programmes and manuals and implementation of pertinent procedures; providing designation of technology and technical support for negotiating its transfer; and carrying out calculations of seismicity and dynamics. (6) NUCLEBRAS sponsored several specialised courses, with the main emphasis on quality assurance, welding techniques and non-destructive testing. (7) Alongside the programme to build up the domestic industry, the government signed a Market Guarantee Protocol with three major industries aimed at ensuring an adequate level of demand for the volume of supplies needed to justify the necessary investments. With respect to non-standard supplies, the government allowed the costs of the technology transfer to be included in the price of the equipment. (8) NUCLEN promoted new production lines for semi-manufactured materials and components. In the case of austenitic and non-ferrous products, it established an extensive qualification programme for forming, heat treatment and welding. It also set up a rigid control system for several production steps, including intensive use of non-destructive testing. All these measures were integrated into the quality assurance system. In addition, the suppliers of ferritic steel for tubes, pressure vessels, heat exchangers, tanks and welded structures, and the suppliers of ferritic steel forgings for nozzles, flanges, connections and large forged pieces, received special attention. Finally, NUCLEN promoted the development of austenitic steel, especially forged steel with niobium and titanium. (9) NUCLEP, the integrated factory for the production of heavy NSSS components, was created. Together with other firms, it should eventually take over production of the turbo-generator set.

The results

Brazil has accomplished a great deal in the fourteen years since the agreement on peaceful use of energy was signed with the FRG. In the case of *nuclear power plant technology*, NUCLEN has been very successful in the area of technology transfer, including the technology needed by domestic engineering companies and equipment suppliers. To qualify as a receiver of technology, NUCLEN staff had to complete systematic training programmes in the FRG and Brazil, with the instruction provided by German engineers working in NUCLEN. In the case of the first two power plants under construction – Angra 2 and 3 – the company trained 196 engineers on the job in the FRG over a period of twelve years. This training load was equal to 2600 man-months. As a result of this training effort, in the area of engineering 77 per cent of the design man-hours are now being handled in Brazil, of which 29 per cent are subcontracted out to private companies. Of the 113 systems in the plants, NUCLEN has taken 108 over from KWU.

In the area of industrial development, 400 of the 800 Brazilian companies registered in the fields of engineering and services, electrical and mechanical equipment and components, instrumentation and control, and materials have been prequalified. Fifty companies have signed supply contracts worth US$500 000 000. Brazilian and West German companies have also signed more than 30 contracts for technology transfer. With respect to electromechanical components, 36 per cent (as measured by value) were produced domestically. Overall, 65 per cent of the value of all the engineering, assembly and civil construction work was handled by Brazilian firms. In terms of the production of heavy components, NUCLEP, with a covered area of 85 000 sq m and a lifting capacity of up to 600 tons, has been in operation since 1980. It is connected to a harbour and has a loading capacity of 1000 tons on roll-on-roll-off ships. Its highly qualified personnel were trained in the manufacture of nuclear components in the FRG and Brazil. It also has its own training centre, where it offers practical qualification in production, quality control, welding and non-destructive testing. NUCLEP is now qualified for the whole range of NSSS heavy components, as well as for the production of components and the provision of services for a wide range of conventional industries. As to the training of plant operators, NUCLEBRAS has installed a simulator that fully reproduces the operating conditions of Angra 2 and 3; its staff of instructors was trained in Germany. This facility has been used since 1983 to train plant operators not only for Brazilian plants, but also for six German plants, one Spanish one (Trillo) and one Argentinian one (Atucha).

With respect to the *fuel cycle*, as a result of prospecting and research begun in 1975, estimated reserves of 301 490 tons of U_3O_8 have been identified, an amount that puts Brazil in fifth place world-wide. A yellow cake plant with a capacity of 500 tons a year was installed near one of the uranium mines. The government is presently commissioning an enrichment plant, NUCLEI, that will use the jet nozzle process. It comprises the first test power cascade, which will in turn determine the parameters for industrial operations. A separation elements plant (FEI) is in operation at the same location. The aim is to reach self-sufficiency in the manufacture of the basic components for the enrichment plant. Finally, Brazil has begun installation of a fuel fabrication plant (FEC), to be completed in three stages: (1) fabrication of fuel rods and structural components for the fuel elements and their assembly; (2) fabrication of uranium pellets from uranium dioxide powder; and (3) reconversion of UF_6 into UO_2. Phase one is already in operation, with a capacity of 100 tons of uranium a year, and FEC has already supplied part of the first two reloads for Angra 1. Phase 2 is planned for 1992.

Present Stage of Development

Difficulties in implementing the programme
Brazil designed its nuclear power plant programme in the economic context of the 1970s, a decade characterized by fast economic expansion and a related increase in energy needs. At the same time, the programme aimed to create an adequate market for a domestic industry that would allow a corresponding technology transfer.

As the economy of Brazil deteriorated in the 1980s, the nuclear programme was affected along with the rest of the country. During the economic crisis of 1983, a time at which two power plants – Angra 2 and 3 – were under construction and two more – Iguape 1 and 2 – were in the design phase and initial stage of site development, the programme had to be drastically reformulated in keeping with the economic austerity. Angra 2 and 3 experienced big reductions in their budgets, and their schedules were set back. Work on Iguape 1 and 2 was postponed indefinitely, a step that interrupted the production of the heavy components, which had already started. NUCLEN and NUCLEP both had to lay off a large number of employees. Completion of the fuel cycle was likewise delayed. Since 1983, the flow of financing has been cut back further and has been irregular. As a result, there have been further delays at Angra 2 and 3, and their start-up dates have been moved back to 1995 and 1997 respectively.

Beyond the effects described above, the reduced budgets and delays in the work have meant higher interest costs on the construction and hence higher overall costs, a lowering of morale of the domestic suppliers of equipment and services because of the delays in delivery and payments, problems in meeting the salaries of the highly qualified group of trained personnel and a related loss in acquired know-how, and delays in the transfer of technology.

Development of an autonomous nuclear programme
Several factors led to the goal of developing an autonomous nuclear energy programme. The jet nozzle enrichment process acquired under agreement with FRG has not yet been tested on an industrial scale. International restrictions precluded the export of the ultracentrifuges used in West Germany to enrich the uranium for its PWRs, as well as other processes. In 1978 the United States interrupted the supply of fuel to Angra 1 because Brazil did not accept the new conditions the US Congress had imposed through the Nuclear Non-Proliferation Act.

A second factor was the financial difficulties, which hindered the intended technology transfer from abroad. A third was that negotiations with France on the construction of a UF_6 plant were interrupted because of conditions unacceptable to Brazil. Fourth, Brazil was unable to buy in the international market the basic materials needed to produce radio-isotopes for medical use. Finally, Brazil was also unable to buy from abroad the uranium enriched to higher levels needed for the research reactors used in the production of the radio-isotopes, even under agreements subject to safeguards.

The autonomous programme, being implemented with the participation of universities and several research centres in Brazil and co-ordinated by CNEN, has had as its main objective to pursue activities that further completion of the nuclear fuel cycle, such as: the design and construction of reactors; the enrichment of uranium for power production, research and radio-isotope production reactors; the production of radio-isotopes for use in medical diagnoses and treatment; and the development of facilities for the irradiation of food and for the fabrication of electronic equipment for medical use in reactor and environmental control.

Results under the autonomous programme
A number of important accomplishments have been achieved under the autonomous programme. Brazil is now producing UF_6 in sufficient quantities for research. A totally domestic research reactor has been designed. Radiological instruments have been calibrated. Facilities for the irradiation of food have been designed. Nuclear techniques have been applied in analysing soils, raising the efficiency of fertilizers, creating genetic changes in vegetables and evaluating agricultural productivity. New materials have been obtained as by-products of the nuclear research, such as

superconductive ceramics, teflon and metal alloys. Finally, uranium was enriched on a pilot scale using the ultracentrifuge method.

Reformulation of the organizational structure
In August 1988 the present government decided, on the basis of a thorough analysis of the programme by two official evaluation commissions, to reorganize nuclear activities in Brazil. Among the changes it made was to close NUCLEBRAS and assign its responsibilities to CNEN and ELETROBRAS, which is in charge of overall planning and control of the production, transmission and distribution of electricity. CNEN is now to handle all activities related to the nuclear fuel cycle deriving from the agreement with the FRG, as well as all research and development being carried out by Brazilian nuclear research institutes. ELETROBRAS is in charge of all activities related to the design, construction and commissioning of nuclear power plants. To this end, NUCLEN was placed under it. These changes have not affected the nuclear agreement with the FRG, which is still in force.

NUCLEAR COOPERATION BETWEEN BRAZIL AND ARGENTINA

The Basis for Cooperation

The fact that each country has chosen a different type of power reactor – Argentina going with natural uranium and heavy water and Brazil enriched uranium and light water – poses no obstacle to cooperation. In fact, the agreement on scientific and material exchanges between the two countries was expanded by the agreement on nuclear cooperation signed in May 1980. Argentina has already supplied zircaloy and yellow cake to Brazil, while NUCLEP made the lower part of the reactor pressure vessel for the Atucha II power plant in Argentina.

This nuclear cooperation, which can contribute to faster nuclear development in both countries, was strengthened by the decision of Presidents Raul Alfonsín and José Sarney to promote economic integration between the two countries. In November 1985 the two men signed a joint declaration creating a working group composed of members from both countries' chancellories and energy commissions, its role being to establish the basis for nuclear cooperation. This group was also to look into the establishment of mechanisms that would allow for the verification of the peaceful intentions of both nuclear programmes. However, the mutual trust that quickly developed as a result of the close cooperation has proven more effective than any verification instrument in arriving at a common understanding.

Priority Areas for Cooperation

In July 1986 both countries signed Protocol No. 11 on mutual assistance and information exchanges in the case of nuclear accidents or radiological emergencies. The agreement occurred at the time of the Chernobyl accident in the Soviet Union, an event that illustrated the need for greater international cooperation on nuclear safety.

In December 1986 the two countries signed an annex to Protocol No. 11 that specifically called for nuclear cooperation, with corresponding objectives and methods of implementation in several areas: planning for nuclear emergencies; analysis of safety issues; protection from radiation; metrology of ionizing radiation; norms for radio-protection and nuclear safety; treatment of people affected by radiation and contamination; procedures for the licensing of nuclear plants and plants that handle radioactive materials and for the transport of radioactive material; licensing of personnel for nuclear plants and plants that handle radioactive materials; quality assurance and regulatory inspections; and management of highly active radioactive waste.

Protocol No. 17 was signed at the same time. It specified nuclear cooperation in the areas of: high density fuel elements for research reactors; nuclear instrumentation, detectors and electronics; enrichment of stable isotopes; research on nuclear and plasma physics; safeguard techniques; feasibility study of a demonstration fast breeder reactor; and non-destructive testing.

Participation of the Industrial Sector in this Cooperation

The Argentinian industrial sector initiated nuclear cooperation between the two countries by the private sector within the framework of the protocols. In the second half of 1986 it made several contacts with the corresponding Brazilian sector. The first binational meeting took place in November 1986, with forty representatives of Brazilian companies meeting with fifteen from Argentina at NUCLEP. The participants reached several accords. They would create a binational Committee of Argentine and Brazilian Businessmen in the Nuclear Area (CEABAN) for industrial cooperation in the nuclear area to represent the business sectors. Its main role would be to co-ordinate integration, define the primary steps to be taken and nominate representatives for working groups that would study each particular case.

As to nuclear plants, the main topics for further investigation included the integration of technical equipment, a commercial agreement for supplying third countries and political support to enforce the resulting actions.The following themes emerged at subsequent meetings. One was the organization of working groups in both countries to integrate three main areas related to nuclear power plants: nuclear industrial capacity (specifically, identification of all qualified suppliers and products in the two countries); requirements in both countries related to licensing and quality assurance; and financing mechanisms for exports in both directions. Second was the elaboration of ground rules for the integration of business policy in the nuclear area. Third was the integration of the results from this committee with the efforts of the two governments at cooperation.

In August 1987 a second plenary meeting was held at Atucha I in Argentina, attended by representatives of the major involved companies. The participants approved the ground rules for the integration of business policy in the nuclear area. The main objectives of the agreement were to increase the binational production of nuclear goods and services, to optimize their quality and reduce their costs and to increase the rate of nuclear development and decision-making in both countries. Looking at these objectives in more detail and classifying them into three categories, they are:

- Economic – increase in the volume of supplies; reduction in the costs of production and services; and earning of foreign currency.

- Technological – increase and upgrading of the technological capacity in both countries; optimization of industrial capacity; full employment of qualified manpower; and competitiveness at an international level.

- Political – integration of these objectives with the official programmes in both countries, independently or together, and developing third world markets in the nuclear area.

The committee also approved a proposal for binational supply for Atucha II and Angra 2. This agreement came about as a result of the efforts of the working groups. It envisaged the export of Brazilian components to Atucha II and of Argentinian components to Angra 2, with financing by third parties. (The details of this area of cooperation will be explained later by Bernal Castro.) The committee on industrial cooperation has been working on this last point for the last two years, trying to find new areas for cooperation under the guidelines of the approved policy.

These efforts at cooperation have resulted in several advances. All qualified companies, supplies and services in the nuclear area in the two countries have been identified, as have all norms relating to quality assurance. As such, exports of supplies and services between the two countries are now possible. The supply criteria for components to be exported to Atucha II and Angra 2 have been defined. These steps culminated in the signing of Annex I of Protocol No. 17 in August 1989 in Brasilia, which authorized the export of components for Atucha II and Angra 2 under special conditions.

CEABAN is now studying other initiatives: an exchange of engineering services in the areas of design, construction and maintenance of nuclear facilities and radiation protection at operating units; the fuel cycle, particularly the possibilities for the exchange of goods and services, the production of zircaloy tubes for Brazilian plants; and the plan of Furnas, which is responsible for operating Angra 1, 2 and 3, to sign a protocol for technical cooperation with ENACE to provide services to Angra 1.

Since CEABAN has been asked to participate in all the meetings of the Permanent Commission for Nuclear Cooperation between Argentina and Brazil, it has reported its activities to the Commission, which in turn has integrated that work into the decisions taken by the chancellories and energy commissions of both countries. This form of cooperation has proven fully satisfactory.

In addition to these co-operative efforts by the business sectors in the two countries, in January 1987 NUCLEN and ENACE, which are responsible for the construction of nuclear power plants in their respective countries, signed a mutual assistance contract. It has enabled these companies to exchange services using acquired know-how for their own field.

NUCLEAR COOPERATION IN LATIN AMERICA

Development of the Nuclear Area in Latin American Countries

In analysing the possibilities for cooperation among the countries of Latin America, it is useful to think of them in two groups based on their degree of development in the use of nuclear energy:

- Those that have begun installing nuclear electrical generation programmes – Argentina, Brazil, Cuba and Mexico. These countries have around 7000 MWe of nuclear capacity that is installed or under construction, equivalent to ten nuclear power plants.

• Those that have achieved different degrees of development in the use of radio-isotopes and nuclear radiation – Chile, Colombia, Peru, Uruguay and Venezuela.

Some characteristics of the nuclear programmes of these other countries are provided below.

Mexico

Mexico has abundant domestic energy resources, especially hydrocarbons, which account for 90 per cent of national consumption. Nevertheless, its hydro, coal and geothermal resources are insufficient in themselves to meet Mexico's energy demand. Natural gas and fuel oil, whose value is high, are expected to be reserved in the long run for uses other than electrical generation. In light of these facts, development of nuclear energy seems to be the logical way to go.

Mexico's first national plant was Laguna Verde, which consists of two units of 654 MW each. The first unit is now linked to the electricity grid, following several delays caused by public opposition; this unit is expected to be a key source of power from 1990 onwards.

Mexico has developed a competent group of over 500 engineers qualified in engineering and plant construction and in the operation of an environmental engineering laboratory that links the nuclear plant with the surrounding environment. Mexico has also installed a plant simulator for training operators.

Cuba

Cuba is constructing two 440 MW PWR-type plants of Soviet design in Cienfuegos. The first unit was expected to go commercial in 1989. An identical plant is planned for the region of Holguin.

Colombia

Colombia is not planning to build any nuclear power plants in the next few decades. However, there is some use, principally by the health sector, of radioactive materials and related equipment to generate ionizing radiation. The health sector employs seven electron accelerators, twenty-eight cobalt therapy sources, twenty-seven gamma scanners and more than 7000 x-ray devices. Imports of radioactive material for medical uses amount to about US$700 000 per year. Around 200 industries use nuclear technology for several processes, such as production control, quality control, and prospecting for and exploration of minerals. The Institute of Nuclear Affairs has a 20 KW research reactor. Currently, negotiations on the installation of a 1 MW reactor and laboratories for the production of radio-isotopes are being completed.

Peru

The strategy underlying Peru's nuclear programme is to develop a basic infrastructure oriented toward the promotion and integration of nuclear technology and national development and the study, evaluation and exploration of uranium resources. Peru has made significant advances in the following areas. With respect to *physical and human infrastructure*, specialized personnel have been trained, and a nuclear research centre, consisting of a 10 MW reactor, auxiliary laboratories, a production plant for radio-isotopes, a National Centre of Radiologic Protection, a Biology and Nuclear Medicine Centre, a Graduate Centre of Nuclear Studies, laboratories and workshops, is almost completed. The centre has 200 professionals involved with nuclear research and the production of radio-isotopes.

As to the *evaluation of uranium resources*, based on surveys of the Macusani Basin, the probable reserves are 10 000 tons of U_3O_8, 30,000 tons as possible reserves and 200 000 tons as potential reserves. In the area of *nuclear applications* the main ones have been in the fields of biomedical science, agriculture and cattle-breeding, and, within industry, radio-sterilization and food conservation.

The long-run plan is to: produce 500 tons of uranium a year using domestic and foreign investments; increase production and productivity in the agricultural and cattle-breeding sectors, with an emphasis on solving the problems of basic food production; and apply nuclear technology to industrial purposes, specifically the use of gamma rays for irradiation for food conservation, radio-sterilisation of materials, and applications in mineral prospecting, among other industrial uses.

While Peru also has important hydroelectrical resources, an unfavourable hydrology and seismic conditions are obstacles to their exploitation. Nevertheless, more intensive use of thermo-generation is possible within twenty years. Other possibilities are geothermal energy, coal and natural gas, and nuclear energy.

Uruguay

Uruguay's main nuclear activities have focused on research and development at the Nuclear Research Centre and on nuclear medicine. A 10 KW research reactor was installed at the centre to train personnel, to provide experience with nuclear materials and work on reactor physics and nuclear physics, and to further the application of nuclear analytical techniques.

Activities related to the production of radio-medicines for use in nuclear medicine are quite advanced, with TC-99M the most widely used in diagnosis. In agriculture the programmes include: phosphorous absorption, photosynthetic activity, effectiveness of fertilizers and veterinary applications. Within the industrial sector the main initiatives have been the development of non-destructive assays and related training of personnel. Around the year 2000 Uruguay may depend on thermoelectrical resources for a significant part of its electrical generation. At the same time, the nuclear alternative may be of great interest, especially in the case of smaller size plants.

Other countries

Reference should be made to Chile, where studies made by the National Energy Commission indicate that a 1200 MW nuclear power plant is possible for the period 2000/2005, complemented by coal-fired plants.

In Brazil, according to ELETROBRAS' plan to the year 2010, present capacity of hydro and thermal plants is to increase from the current 50 GW to 160 GW. At that time, the share of hydroelectric plants should be about the same as it is today – around 90 per cent of installed capacity. Nuclear energy will be produced by six 1 300 MWe plants, including Angra 2 and 3. This plan assumes that huge blocks of energy can be transmitted from the hydroelectric plants in the Amazon region to the main consumption centres up to 2500 km away. If that assumption proves untenable, then more nuclear power plants may be necessary.

Previous Efforts at Nuclear Cooperation Involving Brazil

Up to now few efforts have been made to encourage nuclear cooperation in Latin America. They include those made by the Latin American Section of the American Nuclear Society, such as:

- Publication of a Latin American catalogue of the characteristics of companies in Argentina, Brazil and Mexico qualified to supply nuclear equipment and services

- Creation of a committee, with representatives from Argentina, Mexico and Brazil, to promote exchanges of information on norms, standards and other pertinent measures used in each country's nuclear industry and to foster uniformity of criteria that could lead to regional trade in goods and services

- Sponsoring of yearly conferences, attended by people from member countries, on different subjects in the nuclear area.

Future Trends for Nuclear Cooperation in Latin America

Argentine-Brazilian cooperation – A beginning in Latin America
In my opinion, the examples of Argentine-Brazilian cooperation suggest some possibilities with respect to the rest of Latin America. In the *political arena*, Argentina and Brazil have always been natural competitors for the leadership of nuclear development in the region. Both started their nuclear programmes with the aim of developing national self-sufficiency in engineering services and supply of equipment to the extent possible, even while relying on foreign technology in the initial stages. During this period, as their external debt increased rapidly, they found themselves lacking the financial resources needed by local enterprises, and they experienced slippage in their effort to achieve a comprehensive national nuclear industry, since further external financing usually means greater foreign supplies. The imbalance in foreign trade experienced by both countries, the corresponding economic difficulties and the verification of several joint or complementary efforts that could be carried out with existing resources led logically to binational integration. The way to create such integration is to establish cooperation at the governmental level, focusing on some basic principles:

- development of mutual trust, especially in sensitive areas such as the nuclear field;

- identification of all areas of common interest, aimed at increasing technological and commercial trade;

- promotion of the participation of the industrial sector, which has been hurt by the economic recession of the last decade, in bilateral exports so as to maintain a commercial balance and avoid foreign imports;

- establishment of a common basis for looking at other Latin American markets.

A decision by a government to pursue any of these steps will depend strongly on national interests and political expediency. They are, however, necessary for true cooperation.

In the area of *joint technology efforts*, the situation in Brazil and Argentina with respect to suppliers of nuclear goods and services was analogous to that in Latin America now. In the last fifteen to twenty years, and starting with the construction of turnkey plants, the two countries invested a lot toward developing a national capacity. Although they have not yet achieved full technological independence, their experience shows that binational participation in future plants is possible, even given different reactor types. Joint efforts in research and development, in which both countries have reached different stages, can also prevent duplication of costs.

Brazilian assistance to other countries
Brazil, Argentina and Mexico all have the capacity to implement nuclear programmes in Latin America, starting with imported technology and then creating the conditions for growing domestic involvement. To do so requires rational construction of nuclear power plants based on the energy needs and financial capacity of each country.

Based on its own experience with nuclear power plants, Brazil could assist other countries in the following areas:

- preparation of nuclear legislation, licensing conditions and regulations;

- elaboration of quality assurance systems for the engineering and other elements of the nuclear industry and implementation of a quality assurance programme, eventually including a quality control programme;

- training of human resources, including in the use of foreign technology;

- supply of services – for site investigations, analysis of specifications, design activities, site infrastructure, construction, promotion of a self-sufficient domestic industry, production and assembly of components and systems, training of operators, fuel management, and supplies for the fuel elements in the various stages of the fuel cycle already available in the country.

A cost-benefit analysis of the participation of national engineering and other industries in a country's nuclear programme will depend on the plan adopted for the construction of nuclear power plants within a certain period. Other factors such as available technology, financial capacity and desire to become autonomous will also affect the choice of strategy.

Prospects for cooperation in Latin America

Regional economic integration has long been an aim of many leaders in Latin America. The formation of the European Economic Community and the US-Canada Free Trade Agreement reinforces the need to establish regional trade as part of an overall outward-oriented strategy emphasizing export growth.

Recent studies by economic research groups show that exports will have to grow in the direction of industrialized country markets. Because of the economic crisis of the 1980s in Latin America, the result of a combination of external events (the oil price increases, the slow growth in the industrialized world and the rise in international interest rates) and domestic errors (over-borrowing and mismanagement of the public sector), progress toward economic development has slowed in several Latin American countries. A key problem has been the increase in foreign debt. While exports to countries such as the United States have risen, almost every country has also experienced a decline in the dollar value of imports, a trend that suggests a fall in the region's ability to finance exports from industrialized countries. A more consistent approach to commercial integration in Latin America will need to rely on growth within the region, which in turn depends on a satisfactory resolution of the debt problem, on changing discriminatory trade practices in the industrialized countries, and on deep structural reforms in Latin American countries themselves to address the macroeconomic instability.

At the same time, it is important to recognize the unique aspects of nuclear cooperation. For one, for many years the industrialized nations have not made advanced nuclear technology available to Latin America. Beyond the non-proliferation arguments that are usually given to support trade restrictions, commercial and political interests are another factor, since nuclear energy is vital for industrial development today. Use of conventional forms of energy, although not fully exploited in most countries, will be increasingly difficult because of environmental restrictions, such as the flooding of large agricultural areas by hydroelectric plants and pollution of the air by conventional plants.

Since the cost of developing indigenous technology is extremely high and time-consuming, it makes sense to co-operate regionally so as to share in new advances and in developing new programmes. Some steps that can be taken in this direction include preferential reductions in tariffs, alternatives to dollar-denominated payments and elimination of other discriminatory measures that inhibit joint programmes and exchanges.

Above all, a decision to establish a Latin American market depends on firm decisions by the governments to do so. Their commitment will in turn motivate the private sector to engage in such an exercise. It should not be forgotten that technological and commercial cooperation with the countries that dominate nuclear technology is indispensable in the years to come if nuclear energy for peaceful purposes is to be developed.

Presentation – *José Bernal Castro*

ARGENTINE NUCLEAR DEVELOPMENT

The Argentine National Commission on Atomic Energy and its Projects

The organization that has governed nuclear energy matters in Argentina since their outset in 1950 has been the Comision Nacional de Energia Atomica, CNEA (Argentine Commission of Atomic Energy), which is directly responsible to the president of the country. This condition and the stability over forty years has had a positive influence on the continuity and coherence of projects and their results.

It is appropriate to analyse briefly the stages of nuclear development in Argentina as covered by CNEA as an example of that process in a peripheral and technologically dependent country. The model is not necessarily applicable elsewhere, but it does offer one example for consideration.

The first stage covers the decade 1950 to 1960. In this stage personnel were trained and the first research and development groups were formed. The relationship with the main nuclear centres of the world was intense, stemming from the clearly stated goal of using atomic energy for peaceful purposes. The first experimental reactor of 100 Kw was built, using fuel produced locally.

The second stage, which encompassed the 1960s, initiated production in the following areas: (1) mining and production of uranium concentrate; (2) production and use of radio-isotopes for medicine and industry; and (3) construction of a reactor for radiation with its corresponding fuel elements.

The third stage, which extended through the 1970s, clearly focused on nuclear energy. It involved contracting out the first nuclear plant, Atucha I, of 330 Mw, using natural uranium, heavy water and a pressure vessel. (The feasibility study and contract had actually been completed in the 1960s.)

TABLE 3.1   *Argentine Industrial Participation*

		Atucha I [a]	Embalse [a]	Atucha II [b]
Engineering	(8%)*	0	34	100
Construction	(15%)*	100	100	100
Assembly	(17%)*	53	94	90
Electro-mechan. supplies	(60%)*	15	33	65

* Percentage of total construction costs.
[a] Percentage of Argentine industrial participation in each category.
[b] Yet to be evaluated.

The plant started operation in 1974. Simultaneously, a second nuclear power plant of 600 Mw was contracted out: 'Embalse'. It had an important design difference with respect to the first one. In effect, a system of pressure tubes was chosen (CANDU) instead of pressure vessel, but the same type of fuel, cooling and moderator element – natural uranium and heavy water – were used. This plant went on-line in 1984.

It is necessary to point out a significant difference in the type of contract signed for both power plants. Atucha I was to be completed and handed over in operating condition. On the contrary, Embalse contemplated an important participation of Argentine industry in its construction under the supervision of CNEA. It therefore marked a great step forward by the Argentine industry and engineering sectors in the supply of goods and services.

During the fourth stage, which covered the decade of the 1980s, Argentina, as a result of having two nuclear power plants in operation, acquired vast technological and industrial experience that has brought her closer to self-sufficiency in the provision of nuclear energy with respect to both engineering design and the supply of goods and services.

A third nuclear power plant, Atucha II, was contracted out to KWU of Germany. It had the same design as the first one: a pressure vessel, natural uranium and heavy water. In this period Argentina also consolidated the cycle of fuel production and the country learned that uranium enrichment was being handled domestically.

The increasingly difficult economic crisis in the 1980s that was felt throughout Latin America affected the nuclear research projects in Argentina, especially with respect to their continuity, which had been successfully maintained throughout the previous decades.

TABLE 3.2   *Argentine Nuclear Cooperation Agreements*

Country	Year Signed	Country	Year Signed
Bolivia	1970	Brazil	1980/1986/1989
Colombia	1972	Cuba	1986
Chile	1983	Ecuador	1979
Guatemala	1986	Paraguay	1970
Peru	1968	Uruguay	1972
Venezuela	1980		

Engineering and Industry

This evolution in Argentine nuclear activity has a correlation that is just as important as the production of radio-isotopes or electricity, especially with reference to a developing and dependent country. I refer to the expansion of industry and engineering in tandem with that of nuclear enterprises, which can be an important aspect of production and technology if doing so is a goal of national policy, the case in Argentina.

The above description of the four stages of Argentine nuclear development points out the timing and triggering factors in industrial and engineering development. The situation could be applicable to other countries in the region. In fact, industry and engineering played an increasingly important role in the third and fourth stages in connection with Atucha I and Embalse.

Simultaneously, CNEA imposed specific requirements – basic standards on the reliability and safeguarding of nuclear power plants and their components and services. These in turn promoted technological advances in a wide range of companies, which reached an unanticipated level of development as a result of the demand from the nuclear power projects.

The growth described above was gradual. The high quality of the scientists in the first stages of development and the initiatives that followed were the basis for the progress. The local and foreign training of personnel within CNEA – a necessity that holds for any Latin American country even when their experience and objectives may differ – has been the source of the development of basic and applied projects incorporating Argentina's energy production capabilities and corresponding technological advances.

This evolution proceeded in Argentina despite setbacks caused by the chaotic state of the regional economy. Any other Latin American country could follow suit.

Table 3.1 shows the participation of the Argentine industrial sector in percentages in the different nuclear power plant projects. With respect to the construction of Atucha II, incoherent and discontinuous national policies, together with the acute financial crisis in Argentina since the beginning of the decade, have set the construction of the plant back considerably and have meant that certain components that could have been produced domestically have had to be purchased elsewhere.

LATIN AMERICAN NUCLEAR DEVELOPMENT AND ITS RELATIONSHIP TO ARGENTINA

Nuclear Development in the Region

Nuclear development in Latin America varies by country as a result of their different stages of economic and technological development, internal policies and foreign relations and of their electrical energy characteristics. Some have nuclear energy programmes in the first stages of development, in which concrete technological applications do not extend beyond the use of radio-isotopes. Others have nuclear power stations that generate electricity with strong national participation in their operation and supply. Still others have not pursued nuclear development because their governments are indifferent to or have rejected nuclear energy as an alternative source of electricity.

At present, only three countries have actually built nuclear power plants: Argentina, Brazil and Mexico (the latter in Laguna Verde). Even these three countries are facing financial crises that, especially in Argentina and Brazil, have almost paralyzed their respective nuclear electricity programmes. Therefore the future of nuclear energy in the region does not seem very promising. This fact should not, however, be used as an excuse to abandon nuclear development, particularly where the intent is to use it to generate electricity, as well as to provide a catalyst for greater development of the technical and productive capacity. As to the integration of countries around nuclear growth, it should stem from developments in each country. That is, it should be based on their individual capabilities together with those of the other Latin American countries.

The ever-widening 'technological chasm' separating the countries of the region from the industrialized ones calls for a decision to pursue regional cooperation. Because of the technological differences, capturing markets in the developed world would require a scale of production beyond the reach of individual Latin American countries.

Each time a new technology is to be introduced into the region, we must analyze its compatibility with the situation in the region. Direct transfer of technology from developed countries increases dependence, unless the technologies are clearly identified. The only way to neutralize this situation is to assimilate technologies that can easily be integrated into the region.

Where do these technologies exist? They are found in the countries of the region in advanced stages of development. The issue is their willingness to transfer their technology horizontally, within a wider policy of regional integration.

Nuclear Agreements of Argentina with the Rest of Latin America

In keeping with the concept of *rapprochement*, cooperation and technology transfer, Argentina has signed agreements and conventions with the majority of countries of Latin America at different times since the creation of CNEA. Once it made the decision to continue with its national nuclear programme, Argentina also vowed to make it useful to all of the region and to transfer any acquired experience unconditionally.

In all cases the conventions have been signed with countries whose main concern is 'cooperation on the peaceful use of nuclear energy'. In the case of a few countries these agreements or conventions were signed at a governmental level, in others at the level of the atomic energy commissions. Argentina has entered into agreements on nuclear energy with the following countries; see Table 3.2 (Peru and Brazil will be analyzed separately).

In general terms, the transfer of knowledge has been Argentina's major contribution to Latin America on nuclear issues. This transfer has taken place through international organizations (IAEA, Comision Interamericana de Energia Nuclear, CIEN and so on) by means of signed agreements and contracts. Experts from CNEA have stayed at the nuclear power plants of those countries, while professionals from those countries have done so at Argentine nuclear facilities, participating in courses, operations and projects in both research and development, as well as in the application of nuclear energy, and from the production of radio-isotopes to electric energy. The survey for uranium ore has been of ongoing interest in the transfer of technology.

During the 1980s the main agreements Argentina has signed with other Latin American countries have involved Brazil, Chile, Colombia, Cuba and Uruguay. With the latter, there have been active exchanges in the field of radio-isotopes, and Argentina co-operated on the supply of equipment, the mounting and the putting into operation of an experimental reactor.

This cooperation is continuing with joint studies on the development of a new nuclear plant in Uruguay and feasibility studies for the installation of a nuclear power plant in the near future. Cooperation with Cuba has grown within the last five years, oriented basically toward the field of radio-isotopes.

Agreements Involving Participation by the Engineering and Industry Sectors in Peru

Because of the special circumstances and scope of the supply, it is useful to analyse separately the project for a nuclear research centre in Huarangal, Peru, 30 km from Lima. This centre has a 10 Mw reactor for the production of radio-isotopes, as well as laboratories and auxiliary installations that insure maximum efficiency. Attached to the reactor is a plant that produces radio-isotopes in sufficient quantity to meet Peru's internal demand. This facility is connected with treatment and storage systems for radioactive residues.

Based on a contract entered into by the Instituto Peruano de la Energia (Peruvian Energy Institute) and the Comision Nacional de Energia Atomica, CNEA (Argentine Atomic Energy Commission) in 1978, CNEA was responsible for the design. Argentinean companies provided electromechanical and electronic supplies, including the engineering design and detail, while Peruvian companies undertook the construction including detailed engineering. CNEA directed the construction project in general.

Here is a concrete example of Latin American cooperation along the lines I referred to earlier. In effect, the Argentine proposal had to compete with proposals from countries outside the region, from industrial countries. Peru made a political decision to contract with a country of the region that had demonstrated capacity and that used mechanisms for the transfer of technology that allowed Peruvian personnel to participate in constructing the power plant. This example proves that the limits on integration and cooperation disappear when there is a will and a criterion of staying in the region.

I chose the Peruvian reactor as an example of regional nuclear cooperation. This experience can, however, be applied by other countries through binational agreements. In this case, as in any similar case, Argentina demonstrated that it is capable of supplying quality, exportable, competitive goods. The Argentinean companies put new facilities in the hands of Peruvian technicians by making them participants in work teams and training them to operate the centre and future projects.

NUCLEAR RELATIONS BETWEEN ARGENTINA AND BRAZIL

The Past

Argentina and Brazil are two of the most developed countries in Latin America. Faced with similar problems when it comes to acquiring knowledge that is the basis of economic development and social well-being, it seems strange that the two did not formalize cooperation or exchange of information until the 1980s.

There are many possible reasons to explain the fact that both countries were far apart. The following are those I consider the most representative:

1. A history that comes from colonial times (the eighteenth century) and that gave rise to armed conflict and confrontation over the countries' sovereignty in certain regions.
2. The traditional rivalry of Argentina and Brazil over the leadership of the rest of this potentially rich subcontinent.
3. The interests of other power centres, which worked to exaggerate the rivalries and false nationalism to their own ends. Certain sectors within each of the two countries generally supported this foreign manipulation. These three factors apply also to the field of nuclear development.

A fourth reason is often mentioned: until the mid-1970s Argentina was more advanced than Brazil in the nuclear field, a fact that contributed to their governments' failure to enter into agreements. Even though this difference in their level of nuclear development may hinder integration, it should not prevent working toward that goal. Further, in my opinion this difference is not necessarily a disadvantage, especially if the political will to overcome it exists, backed by clear plans for a transfer of knowledge – as in the nuclear case, for example – and the safeguarding of mutual rights. If the hypothesis of the effect of an inequality in development is valid, then the rest of Latin America will not be able to reach any agreement with Argentina and Brazil. This region would be condemned to nuclear fragmentation, and the development of these two countries would disappear.

Nuclear Agreements

1980
There are two reasons for the radical change in the 1980s with respect

to nuclear policy issues. The first is internal, while the second has an important foreign component. With respect to the first, both Argentina and Brazil came to view the conflict differently: it was seen as hurting the interests of both nations by suppressing cooperation as a genuine expression of regional interest.

The second factor is the international pressure that emanates from the view that Argentina and Brazil are locked in an arms race. Indirectly this view induces both countries to seek agreements so as to counter this view. Cooperation is the antithesis to the artificial confrontation promoted from outside.

In this context, in 1980 the two governments signed their first agreements on nuclear issues. They can be summarized as follows:

1. The atomic energy commissions of both countries were to be in charge of signing and undertaking agreements on nuclear energy.
2. The following agreements and conventions were entered into:
   (a) Agreement on cooperation between the governments for the development and application of the use of nuclear energy for peaceful purposes.
   (b) Convention on cooperation signed by the atomic energy commissions that refers mainly to the field of radio-isotopes and nuclear safeguards. The following protocols were based on this agreement: (i) in reference to the training of personnel – its goals were not met; (ii) in reference to technical information – it did not meet expectations.
   (c) Agreement on cooperation between CNEA and Brazilian Nuclear Enterprises, or NUCLEBRAS (Empresas Nucleares Brasilenas), to involve applied research and technology and oriented toward the generation of nuclear power. The following protocols were based on this agreement: (i) Acquisition of 160 000 m of zirconium alloy tubes for NUCLEBRAS for the manufacture of fuels destined for Brazilian nuclear power plants. This trade has virtually not taken place yet. To date only 1000 m in order to qualify the Argentine facilities have been delivered. The contract is being renegotiated; (ii) Loan of 240 tons of uranium concentrate to NUCLEBRAS; (iii) Transfer by CNEA to NUCLEBRAS of the technology of lixiviation of uranium ore. This contract was never signed because of the characteristics of Brazilian uranium ore; (iv) Participation of NUCLEBRAS in the construction of the Atucha II power reactor. NUCLEBRAS did manufacture the lower section of the pressure vessel of Atucha II.

The above agreements signed in 1980 have, as is clear, been only partially completed, while part of the proposals are still suspended. Nevertheless, they were clearly instrumental in changing the status quo that had prevailed until then.

From 1980 on, the sharing of nuclear activities increased and was promoted. While the development of technology and applied science may not have had the desired continuity, it did help end the foreign political pressure that with the excuse of the arms race harassed both countries.

1985/1986

This stage is initiated with the meeting of Presidents Alfonsín and Sarney in November 1985 which led to a Joint Declaration on Nuclear Policies with the following common goals:

1. Development of nuclear energy exclusively for peaceful purposes.
2. Cooperation on all aspects of the use of nuclear energy for peaceful purposes.
3. Extension of this cooperation to all Latin American countries with similar goals.

A task force was formed under the leadership of both chancellories, with representatives of those institutions and of the nuclear energy organizations. In this manner, the will to integrate and complement the science and technology of both countries was incorporated into the political goal.

Three subgroups of the task force were formed with the following characteristics and goals:

1. To address the political issues in the international situations of both countries.
2. To work on the technical issues involved in co-ordinating the scientific and technological aspects.
3. To resolve the political and legal problems arising from cooperation between both countries.

The task force has taken concrete actions, as demonstrated by the signing of the two protocols that make up the agreements on economic cooperation for Argentine-Brazilian integration:

1. Protocol No. 11, 'Protocol on immediate information and reciprocal assistance in the case of nuclear accidents and radiological

emergencies,' applied specifically in the unfortunate accident at Goiania, Brazil.
2. Protocol No. 17, 'Protocol on nuclear cooperation', of which the following aspects should be noted:
   (a) Development of low enrichment fuel elements for research reactors.
   (b) Development and exchange of nuclear instrumentation.
   (c) Research and cooperation on nuclear fusion.
   (d) Joint projects on radiological protection and safeguards.
   (e) Joint project for the development of technologies that lead to the construction of fast breeder reactors.

In general, the goals of these protocols had been met as of 1986, the date they were signed, in the face of the obvious difficulties of all new processes as well as the tough economic situation of both countries.

Political Aspects

The governments of Argentina and Brazil have proven that cooperation and integration are possible around nuclear energy technology. Presidents Alfonsín and Sarney have referred to this concept since 1986 in their periodic visits to the most advanced installations of nuclear technology in both countries, such as the uranium enrichment plants in Pilcaniyeu, Argentina and Ipero, Brazil. Precisely in this last city in April 1988 a Presidential Agreement made the task force a 'Permanent Committee' with more extensive and specific functions. This committee now regulates integration and cooperation related to nuclear activities between both countries.

Nuclear Industries: CEABAN and the Extension of Protocol No. 17

As a consequence of the needs of their nuclear power plants and of their national policies, Argentina and Brazil have developed industries for the supply of nuclear goods. The *rapprochement* initiated in 1980 and promoted in 1986 stimulated the nuclear industrial sectors in both countries to hold a number of meetings that finally led to the creation of the Argentine-Brazilian Nuclear Industry Committee (CEABAN) in 1987. The initiative came from the Argentine Industrial Group of the Argentine Association of Nuclear Technology (AANT).

The goal of integration that both the Argentine and Brazilian business organizations were pursuing involved complementing the capacities and optimizing the technologies so that this sector would become increasingly independent of foreign intervention. This effort prompted an extension of Protocol No. 17 that CEABAN first presented to both Chancellories in mid-1988. The three reasons for seeking a modification of Protocol No. 17 were the following:

1. The need to have an official document that referred to the contribution of the nuclear industrial sectors in both countries to integration and cooperation.
2. The need to define a concrete plan for putting the intent of integration and cooperation into practice relative to the export of nuclear products between Argentina and Brazil.
3. The need to have financial aid from official banks for commercial operations.

The final text of the extension of Protocol No. 17 was discussed for several months by the business representatives, Chancellories and atomic energy commissions. Then in August 1989 the Chancellors of both countries approved the final version and signed it in the presence of their respective Presidents.

In this manner, a chapter in the history of Latin American nuclear development was completed and a new one opened, in which the private sector is beginning to provide nuclear supplies alongside the official governmental sector. For the first time in the history of this part of the continent the private nuclear industrial sectors of both countries agreed to participate, even as competitors, in the two most important nuclear projects being undertaken at that time, Atucha II and Angra 2.

The new protocol included a list of similar items at both nuclear power plants whose value came to approximately US$30 million. This original list was issued to qualified businesses in both countries so that they could offer their respective bids.

The financial problem that has stalled continuation of the construction of the two plants has not been solved in spite of the US$30 million included in the Protocol. However, the main purpose of the extension of Protocol No. 17 was to demonstrate that the official and industrial sectors in both countries could sign an agreement on nuclear issues, generate a document and complete the necessary actions to promote the integration and cooperation of the productive nuclear capacity in their countries.

The extension was not arrived at easily, nor will its implementation be easy. Historical, political and economic prejudices have to be overcome, within the context of an economic crisis that is increasing the doubts and mistrust between both nations and that makes them tend to prefer to buy their supplies from industrial countries, as they are the only ones with the financial capacity.

Thus, putting the extension of the protocol into practice is quite difficult. On the other hand, closed competition is self-destructive. With respect to the other technological developments Argentina and Brazil wish to implement, as in the case of nuclear development, pursuing them independently will leave the two countries in a weaker position relative to the industrial countries. This problem, which will be increasingly evident, was noted by both Presidents Menem and Sarney at meetings held in Brasilia to sign the protocols.

Finally, in the field of electro-mechanical components at least sixty companies in Brazil and Argentina have a large body of highly qualified professionals and technicians. At present, and despite the stagnant situation in both countries, they have been able to keep their production capacity, engineering and services intact, if not united. If a definite programme of construction of power plants existed, it would be possible to say that the capacity to supply is adequate except for a few parts that are required in small numbers or that have special characteristics. The technical scientific infrastructure is also present, mainly concentrated in the atomic energy commissions, which provide the basic component of the nuclear energy projects. All this technological and productive capital has been called into service by the signed nuclear protocols. Work is continuing to put these protocols to more efficient use.

FINAL CONSIDERATIONS FOR THE FUTURE

The Latin American Economic Substrate

I have analyzed nuclear developments in a country like Argentina, which are clearly transferable to another country like Brazil. I have shown how relations with other countries have been generated. I have looked closely at the Argentine-Brazilian case as an example of nuclear cooperation and as a forerunner of Latin American integration.

I have also described the present state of nuclear developments in the two countries, with specific reference to the economic crisis suffered by Argentina, which has delayed the inauguration of the third nuclear power plant despite the country's large energy deficit (a similar situation exists with the nuclear programme in Brazil).

Faced with this difficult reality, is there reason to hope that protocols such as that signed recently by both countries in the area of nuclear energy could help development, and even extend it, to the rest of the region? A number of points help predict the future, to achieve this the best thing to do is to analyze the immediate past. Maybe from there valid experiences can be derived that will allow us to correct the errors and consolidate the successes.

Latin America is facing an increasingly protectionist policy on the part of the industrial nations. However, non-traditional exports increased during the 1980s, although the paradox of the international economic system does not allow that increase to be converted into development capacity, since the surplus of the commercial balance of payments goes toward the interest on a massive foreign debt.

I will present a few figures on the Argentine economy that will serve as parameters for analyzing the general economic deterioration and its consequences in the nuclear technology field. Since the decade of the 1950s until the mid-1970s, the Argentine economy increased 4 per cent annually, and fixed national investment represented approximately 22 per cent of the gross national product (GNP). In the second half of the decade of the 1980s GNP still had the same absolute value as in the mid-1970s. Production per capita was, however, 20 per cent lower, and domestic fixed investment was 13 per cent of GNP.

These and similar facts too numerous to mention are mainly a consequence of the economic paralysis and foreign debt. Between 1981 and 1986 Argentina paid US$30 000 million in principal and interest, of which US$10 000 million were refinanced by the creditors and the rest paid using the country's own resources. However, Argentina's foreign debt rose in this period by more than US$20 000 million. As a result the debt service took 40 per cent of all export earnings.

The situation in Latin America is clearly reflected in the recently published *Report on the Economic and Social Progress in Latin America* of the International Development Bank in 1989. With reference to the foreign debt, the report emphasizes, among other things, that 'international interest rates have increased constantly, sharply accentuating the service charge'. It notes that Latin America's per capita GNP decreased globally by 1.5 per cent over the previous year. The listing of figures in the report shows a sombre social and economic panorama that would justify a change in the title of the report from *Progress* to *Anti-Progress*.

This situation is the product of irrational public investment. Only a political decision to implement structural change and growth for society can lead to rational public investment and more dynamic private investment, backed by a previously approved coherent plan.

Policy and the Nuclear Future

I believe that the rational criteria creditors have imposed will not generate sufficient growth to satisfy national objectives. Among those objectives are the development of nuclear technology. I do not believe that an independent, integrated nuclear development programme for the region can be part of a proposal by the industrial countries or blocs.

Up to the time being this has been the situation for Argentina's nuclear plan and for the Brazilian information programme, another advanced technology field. A year ago in Buenos Aires the Brazilian Minister of Science and Technology, Deputy Luiz Henrique da Silveira, said:

> The Brazilian market for mini and micro computers is amongst the six largest in the world, according to the US Department of Commerce. In ten years 330 businesses have flourished employing 50,000 workers. The participation of the national businesses in this market, in number values, was 2% became 52% in 1987.

The Minister added,

> This process is not occurring without problems, such as those well known problems which we have had during the last few years with the developed world, which do not want to lose such an increasingly promising market.

Undoubtedly, the effort, creativity and ingenuity to overcome the ongoing recession is in the hands of the countries themselves, the members of the Latin American bloc. Technology as an expression of applied intelligence is one path that they have to walk together. Without doing so they will not emerge from their stagnation. The nuclear field is an obvious component of the process of integration. As President Sarney of Brazil said, 'The international division of power is increasingly in the division of knowledge.'

A concerted project such as the nuclear project of Argentina and Brazil might not immediately have the desired results. The poverty in the two countries does not permit it, much less deduce from the economic component the success or failure. The agreement has simply initiated a mechanism for integration, which, using technology, science and nuclear industry as the goal, will demonstrate that independent and integrated intelligence put to the service of Latin American society is the only rational road to changing the negative slope of the curves.

The concept of integration as a form of growth is not new. It is extensively practised in the industrial countries, which are grouped in clearly defined blocs.

Latin America cannot remain passive. The presidents of Argentina and Brazil during their meeting in Brasilia last August made this point repeatedly. All sectors of their governments should understand that within the framework of integration is the only possible road. In the nuclear area and within the framework of the signed agreements and related actions, representatives of this technology and users of nuclear facilities regularly grouped in the states – together with the private industries that produce basic goods – should make plans that, with a minimum regional financial backing, would permit the development of this activity with the largest possible multinational Latin American component.

Industry, Technological Capacity and the Nuclear Future

Latin American industry, especially that dedicated to the nuclear field, had to undergo a deep transformation to reach the level of technical development required for nuclear safeguards. An integrated growth plan for the region using compatible technologies would allow optimization of the costs and would create the necessary conditions for sustained and extended growth consistent with Latin America's needs.

Some regional Latin American nuclear projects have been adapted to the needs and capacities of the region. These projects were generated using local engineering and were appropriate to the productive level. I am referring to the compact and medium-power nuclear power facilities, experimental reactors, radio-isotope production plants, manufacturers of nuclear fuel, and processes and manufacturing facilities for nuclear supplies. Together with this engineering and production capacity comes adequate development of metallurgy, chemistry, mineralogy, welding, non-destructive testing, quality assurance, applicable general standards, etc.

This scientific, engineering, technological and industrial capacity, integrated and made compatible with the different needs of the region, offers an optimistic panorama if the regional nuclear policies are clearly defined and the economic decisions correspondingly oriented.

Specifically, and with respect to the advanced Argentine-Brazilian agreement, I consider it necessary that its impact grow and expand in the future, that the means of *rapprochement* between similar sectors be accentuated and that mutual knowledge be deepened so that the fears disappear. Consequently, I propose the following as mechanisms of consolidation:

1. A detailed survey of the scientific and technological productive capabilities in each country.
2. Assured access to the binational economic capacity to fund joint nuclear projects.
3. The commitment of both governments to support and encourage initiatives in this field.
4. The commitment of the industrial sector to make available its intellectual and material capacities for balanced integration and cooperation.
5. The commitment of the nuclear client to refer the enterprises of both countries to domestic suppliers of goods whose capacity to provide them has been proven.
6. Integration of local participation through a collegial organization where all sectors are represented and where the agreements reached have a real possibility of being met.
7. Complementing of this plan with the shared policies and legislation necessary for it to be undertaken.
8. Analysis of those aspects of this agreement that will allow it to be extended to the rest of Latin America.

As soon as Latin American integration in the nuclear field starts to become a reality, and not just an expression of desire, oriented toward a single goal – an increase in the well-being of the people and conservation of their respective identities – as soon as we can say that this integration has been attained, the central blocs will have to accept it and will begin to understand that this part of the world also has a right to establish its own criteria for growth.

In this process of understanding and acceptance, a sincere and balanced cooperation could be established between the developed blocs and Latin America, independent of the level of development that separates them.

Response – *Samuel Edlow*

I am not a politician, theoretician or bureaucrat. I am an entrepreneur – a plain, everyday businessman. I think the function of a businessman is as follows: after the politicians and the bureaucrats have established the rules, policies and regulations, we are the ones who implement those decisions and policies. We go forward and do the work.

When it was decided, for example, that physical protection of nuclear materials was imperative and the policies were set, we were the ones – my company, that is – who set up the armoured car service and put those policies into practice in the everyday world. When design criteria were established for a spent fuel shipping cask, we were the ones who first designed and built the cask and did the transporting.

We face a world of harsh realities. The views I express here today come from thirty-five years of facing reality. Why do I tell you this? I tell you this to establish my credentials as a practical man, so that you will know that what I have to say in the next few minutes comes from a very practical point of view.

The scholarly papers that members of this panel have presented detail the development of the business side of nuclear power in Argentina and Brazil. They speak of both success and failure. I think there are no differences of opinion on this among the panel members here today.

When I went to the university, I took a course in public speaking. I was told that the most important thing to do was to define one's terms. The subject of this discussion is, 'What are the industrial and economic benefits of Latin American nuclear cooperation'? Let me address myself first to the word, 'cooperation'.

What comes to mind when one uses the term 'cooperation?' I think of parents and teachers co-operating to build a better school. I think of apartment owners and tenants co-operating for a higher standard of living in an apartment building. I think of parents and children co-operating to provide a better and happier home life. I think of companies co-operating to promote mutual research for better products.

What image comes to mind when we speak of bilateral cooperation in the nuclear field? I believe it depends on the extent of the cooperation and its goals. I am speaking now from an industrial and economic point of view, not about politics or philosophy.

I believe the potential exists for huge industrial and economic benefits from such cooperation. It is my view, however, that there are no significant economic or industrial benefits to be realized from Brazilian-Argentine nuclear cooperation as it is now structured.

Why do I say this? Because the cooperation now is limited to the development of indigenous technology and production facilities, the cost of which is unconscionable, in my opinion, and unjustifiable from an industrial and economic standpoint.

There are undoubtedly tremendous political benefits. There are tremendous philosophical benefits. There may even be mutual military benefits. I do not believe, however, that there are any real industrial and economic benefits as this cooperation is presently structured. It is true that the elimination of duplication of effort – allowing each country to provide that which it can provide best – does afford some savings. However, I believe these savings are relatively unimportant.

Now, let me turn from the negative to the positive. Whether we like it or not, whether it is right or wrong, reality tells us that important technologies and services are not available to Brazil and Argentina because they are unwilling to accept a system of safeguards covering all their nuclear facilities. Mr Fernando Henning makes my point for me when he writes, 'It should not be forgotten that the technological and commercial cooperation with the countries that dominate nuclear technology is indispensable in the years to come if nuclear energy for peaceful purposes is to be developed.'[1]

Now we come to the facts of life. To obtain those technologies, an acceptable safeguards system is essential. Here we find the real industrial and economic advantages of cooperation: it allows each country to stop trying to re-invent the wheel. With safeguards in place, new technologies that will be essential in the future will be available to Argentina and Brazil. In the presence of a safeguards system that is acceptable to the developed countries, it will be possible to bring new technologies to the developing countries – Brazil and Argentina, in this case – so that substantial and real industrial and economic advantages can derive from their cooperation.

Let me quote from Mr José Bernal Castro's paper:

> As soon as Latin American integration in the nuclear field starts to become a reality, and not just an expression of desire, oriented towards a single goal, an increase in the wellbeing of the people and conservation of their respective identities, as soon as we can say that this integration has been attained, the central blocs will have to accept it and will begin to understand that this part of the world also has right to establish its own criteria for growth. In this process of understanding and acceptance, a sincere and balanced cooperation could be established between the blocs, independent of the level of development that separates them.[2]

A satisfactory safeguards system will bring about what Mr Bernal Castro refers to.

Let me conclude by telling you a brief story. I was discussing a policy matter with a very close friend of mine and trying to convince him the policy ought to be changed. I said to him, 'What I am saying is logical.' His answer was, 'Sam, what makes you think governments act logically'?

If the governments will face today's reality, then tremendous economic and industrial benefits can accrue, and logic can eventually prevail.

Discussion – *Dr Walter Cibils*

First, I want to express the pleasure of the nuclear authorities of Uruguay, and my own appreciation, for your having selected our country as the site of this important meeting. Also, speaking exclusively for myself, I would like to express what an honour it is for me to be part of this Table.

I am going to talk about nuclear cooperation in Latin America and in particular the benefit that has been derived from nuclear relations between Argentina and Brazil, which has already been discussed in this room.

Latin America is seeking to improve its position in relation to the world nuclear community as a whole. For a rich person to exist there must be a poor one; for a happy people, an unhappy one. Well-being, happiness and wealth are relative concepts, not absolutes. Therefore the quest for the better place to which we refer is one of hard competition.

Technological development plays a dominant role on this road to development. Nuclear technology is only a tool to attain the goals of development. Argentina and Brazil are in the forefront of this Latin American initiative, having embarked upon this venture with enthusiasm, intelligence and dedication, which have made it possible for them to obtain today's high standard.

Their cooperation should be understood as a means to an end. The prosperity of the countries of the region comes through Latin American integration, of which nuclear integration is a part. Argentina, Brazil and Uruguay are making great efforts to make it possible. We are closely following the efforts of integration of Western Europe and between Canada and the United States. Only through integration can we improve Latin America's position relative to the rest of the world, economically as well as technologically.

Science and technology are closely linked to the goal of global development of nations. In Latin America there is no consensus, however, on the best way to achieve this development. For some it is a matter of following the development path of the industrialized countries. Others, including myself, believe we should design our own strategies suited to particular situations.

We believe it is absolutely valid to use the experience of other countries that have faced similar situations. For those Latin American countries that have not reached Argentina's and Brazil's level of development in nuclear technology, but have been spectators of their effort, this initiative allows them to draw conclusions, avoid the same errors and join a co-operative effort having them as a guide.

Nuclear technology appears to be a feasible path to technological development, and especially important for resolving the energy problems of some countries. Countries such as ours, which have exhausted their hydroelectric capacity for generating electricity, and which do not have natural resources of coal and oil, must look to nuclear energy as a viable solution to their energy problems.

In a framework of regional integration, Uruguay and other less industrially developed countries should make use of the nuclear capacity of Argentina and Brazil through projects with close cooperation. Not having important industrial resources should not be an obstacle to carrying forward this type of enterprise. Success of such enterprises will depend essentially on the human resources used. We are convinced that in our countries human resources are the principal capital.

Nuclear cooperation in the framework of Latin American integration will be possible in the short term and will have Argentina and Brazil as natural leaders. These countries have made great efforts not only in the field of energy, but also in areas such as medicine, agriculture, industry, environmental protection and radiological security. Latin American countries have already advanced in this cooperation since 1985 with the implementation of the ARCAL programme (Regional Co-operative Arrangements for Promotion of Nuclear Science and Technology in Latin America), which supports the International Atomic Energy Agency. We are always alert to the possibility of increasing our knowledge as a result of the experience acquired by the more advanced countries.

We believe that nuclear cooperation, as well as scientific, technological, industrial and economic cooperation, will provide the basis for a regional integration that will permit us to achieve great gains relative to the rest of the world.

4 Nuclear Confidence-Building: Models for a Bilateral Safeguards and Verification Regime

Presentation – *Dr William A. Higinbotham*
and Helen M. Hunt

SAFEGUARDS ARRANGEMENTS FOR A BILATERAL
NUCLEAR CONFIDENCE-BUILDING REGIME

The term 'safeguards' is used in two senses with respect to nuclear materials and facilities. One is the international application of measures to provide mutual assurance that nuclear materials and facilities are being used only for declared peaceful purposes. The other is with reference to national programmes for the control and physical protection of the materials and facilities. Multinational nuclear assurance programmes depend on the existence of these national programmes, as noted below.

The well-known international safeguards system of the International Atomic Energy Agency (IAEA) and the regional safeguards system of the European Community (EURATOM) are reviewed briefly. A number of less formal mutual assurance programmes that have been discussed are also mentioned here. In a companion paper, Dr Milton Hoenig reviews verification techniques that have been or are being implemented to provide assurance that certain multilateral agreements are being observed.

Any multilateral agreement must be voluntary, with each party concluding that the undertaking will contribute to its security and welfare. In the case of the EURATOM countries, the objective was to co-operate in the development and exploitation of nuclear energy as well as to provide safeguards assurance. In the case of the IAEA, a number of nations endorsed it to promote the development of nuclear energy and to ensure that such developments were not exploited for military purposes.

International or multinational safeguards systems must be based on national safeguards systems because the nation-state has the authority to define and enforce regulations for the control, accounting and physical protection of nuclear materials. The existence of an obviously effective and open national system for the control of and accounting for nuclear materials should, of itself, provide a considerable degree of assurance to other nations of peaceful intent. For these reasons it seems logical to begin with a discussion of a national safeguards system. Next we briefly mention some of the features of IAEA safeguards, with which this audience is already familiar, and outline the EURATOM safeguards system, not so much as an example for other states to consider but rather because it may contain some features and represent some experiences that may be relevant. Finally, we mention some suggestions as to less complicated steps that two or more nations might take to provide assurance to each other and to others.

A NATIONAL SAFEGUARDS SYSTEM

The purpose of a national safeguards system is . . . to ensure government control of the possession, use, and production of atomic energy and special nuclear material so directed as to make the maximum contribution to the common defense and security and the national welfare, and to continued assurance of the government's ability to enter into and enforce agreements with nations or groups of nations for the control of special nuclear materials and atomic energy.[1]

As is the case with other programmes of national importance, a government agency, such as an atomic energy commission, is established to implement the government's objectives and policies. This agency may operate the nuclear facilities itself, contract with non-governmental organizations for their operation or license private organizations to possess nuclear materials and operate the facilities. In all cases the government agency must define how the objectives and policies are to be performed and ensure that they are performed satisfactorily. The safeguards activities of material accounting, material control and physical protection must be performed at the facility level and be supervised by the government agency.

Nuclear materials are valuable. Some have important security implications; some are highly radioactive. Accurate and timely accounting of nuclear materials is important for all these reasons. Access to the materials and the facilities is controlled to exclude possible adversaries and to admit selected and authorized employees and supervisors. Since natural and low-enriched uranium are only slightly radioactive and cannot be used for nuclear explosives, traditional physical protection measures are adequate. However, expensive and sophisticated physical protection measures are required for high power nuclear reactors and highly radioactive materials, which might be targets for sabotage, and for facilities and shipments of highly enriched uranium and plutonium. Both Argentina and Brazil have extensive experience with material accounting and with physical protection. The IAEA has published guides for material control and accounting and for physical protection.[2,3] A number of non-nuclear weapon states, including Brazil, described their physical protection systems in an issue of the *Journal of the Institute of Nuclear Materials Management*.[4]

It is useful to say a little here about the experience of the United States in defining and implementing its nuclear policies. One objective of the Atomic Energy Act of 1946 was to provide for civilian control of the development of atomic energy for civilian and military purposes. The act assigned legislative oversight to a joint Senate-House committee and authorized the Atomic Energy Commission (AEC) to monitor its own safety and safeguards operations. In 1954, the act was amended to permit the lease of nuclear materials to private companies and to provide assistance to other nations. However, the Joint Congressional Committee was at a disadvantage in that only a few members of the Senate and the Congress had access to inside information, which considerably restricted public knowledge of and review of the civil and military programmes. The self-policing of the AEC allowed it at times to be rather casual about safety and safeguards. Only after about 100 kilograms of highly-enriched uranium were found in 1966 to be missing from a company fabricating such fuels for the AEC were well-defined requirements for material control and accounting finally initiated and enforced. Then, in 1974 the AEC was split into the Nuclear Regulatory Commission (NRC) and the Energy Research and Development Authority (ERDA, soon to be reconstituted as the Department of Energy). The latter retained responsibility for the military production programme and energy R&D. The NRC became responsible for safety and safeguards at private, licensed nuclear facilities. The Joint Congressional Committee was abolished, and a number of congressional committees are now involved in reviewing the performance of both agencies.

I review this history to emphasize the importance of separating promotional from regulatory activities, of separating the responsibilities for safety and safeguards from those for production, and the importance of openness for public review of policies and programmes.

IAEA SAFEGUARDS

The IAEA and its safeguards activities are well-known to those of you at this meeting. The Agency was established in 1957 to promote atomic energy and was instructed to ensure, to the extent it reasonably could, that nuclear materials and technologies supplied by the Agency or subject to its supervision were not used for military purposes.

The Agency has two major missions. One is to promote nuclear energy: it is to assist member states in obtaining information and training in the uses of nuclear energy and in obtaining materials and equipment. This assistance takes many forms and today is primarily of interest to developing countries. The Agency operates the Trieste theoretical physics institute, an oceanic institute on the Mediterranean, and a chemical laboratory near Vienna. However, typically it works with the requesting and donor countries to arrange for the transfer of materials, experts and other means of assistance. The IAEA's second mission is to define and apply safeguards under the terms of the INFCIRC/66 and INFCIRC/153 agreements.

The IAEA also performs a number of important services for its members, such as maintaining an abstracting service on nuclear-related publications, an international nuclear data file, exchange of information on reactor operations and safety-related incidents, and management of international or general conferences on nuclear or related subjects.

Following ratification of the Nuclear Non-Proliferation Treaty (NPT) by the required number of countries, the United Nations modified the formula for calculating members' dues to reduce the amount owed by developing countries and to increase that owed by countries with larger nuclear programmes. Some countries make additional voluntary contributions or provide technical assistance to the Agency. IAEA's technical assistance and safeguards budgets are not large when compared with, for example, the costs of many national safeguards systems.

The IAEA has published a number of documents that describe what it is and how it operates.[5] Each year it issues a 'safeguards implementation report', which it sends to member states. Some of the information in the report is made public, while some is treated as sensitive.

The Agency's method of drawing conclusions as to the assurance provided each year by its inspection activities is complicated. In any case, each nation has to decide how much assurance the IAEA is providing and the value of the programme to it and others. The safeguards budget has been frozen for the last two years, although the amounts of nuclear materials continue to increase.

Those at this meeting who are in responsible government positions should have access to the confidential annual 'safeguards implementation reports'. They and others may have had the opportunity to participate in the technical and advisory group meetings the Agency has sponsored from time to time. About fifteen years ago the Director General of the IAEA established a Standing Advisory Group on Safeguards Implementation (SAGSI). A representative of Brazil, Fernandez Beanchini, is presently a member. Among other subjects, SAGSI reviews and comments on the Agency's procedures for inspection under the INFCIRC/66 and INFCIRC/153 agreements, a subject that should be of particular interest to non-NPT members.

The participants in this meeting undoubtedly understand the difference between the INFCIRC/66 and INFCIRC/153 agreements. Under the latter the state submits all its nuclear materials to IAEA inspection, and the Agency receives reports on all international transfers, inter-facility transfers and facility inventories. The safeguards procedures include analysis of the reports, auditing of the records maintained at the nuclear facilities and independent verification of some of the quantities transferred and of those residing at the facilities. Since all the materials are reported and available for verification, redundant information exists that supplements the information obtained directly from the documents and the inspections.

In the case of the INFCIRC/66 agreement only some materials are subject to safeguards. INFCIRC/66 safeguards apply both to nuclear facilities and to some non-nuclear materials to compensate for the fact that not all the materials are under safeguards. Nuclear materials that are produced from safeguarded materials or in safeguarded facilities become subject to safeguards. This system makes the design and implementation of safeguards more complicated for the IAEA and sometimes more burdensome for the nation involved. For example, the government of Switzerland requires that the IAEA safeguard the heavy water production plant sold to Argentina. Design of a practical safeguards system for such a plant is quite difficult. It is complicated to track what materials should come under safeguards because they are produced from safeguarded nuclear or non-nuclear materials. If Argentina were to accept IAEA safeguards on all its nuclear materials, it is likely the IAEA would drop the requirement for safeguards on non-nuclear materials and the heavy water plant.

EURATOM SAFEGUARDS

In the early 1950s the Western European countries (France, the Federal Republic of Germany, Belgium, the Netherlands, Italy and Luxembourg), devastated by World War II, tentatively formed the European Community. As a part of this system the participants agreed to form EURATOM in order to develop nuclear energy for the community as a whole. EURATOM has a complicated structure, with an Assembly, a Council, a Commission, a Court of Justice and an Economic and Social Committee. The intent was that the organization would perform or control nuclear energy R&D and that it would have a management role in the development and operation of nuclear power and associated facilities. In effect, the organization was to own all special nuclear materials (enriched uranium, etc.). All contracts with other nations for nuclear materials, technology, etc. were to be arranged through EURATOM or approved by it. There were provisions to ensure that each member state would have equal access to nuclear materials, information, patents and so forth. EURATOM was to establish research laboratories and to co-ordinate the R&D of member states. The main laboratory for the community was established near Ispra, Italy.

A few years later France decided to pull out of the joint R&D effort and to develop its own programmes. Implementation of the European Community also slowed. However, the safeguards element was maintained, as noted below, and the members continued to negotiate sales, purchases and agreements with other countries through EURATOM. For example, the United States has sold uranium and enrichment services to the member states through EURATOM, and EURATOM has approved purchases of technology and patent rights by firms in member states. As the member states ratified the NPT, they conducted the negotiations with the IAEA through EURATOM.

Safeguards are described in section VII of the EURATOM charter as follows:

> In accordance with the provisions of this Chapter, the [EURATOM] Commission shall satisfy itself that in the territories of Member States,
>
> (a) ores, source materials and special fissile materials are not diverted from their intended uses as declared by the users;
>
> (b) the provisions relating to supply and any particular safeguarding obligations assumed by the Community under an agreement concluded with a third State or international organization are complied with.

The commission may send inspectors into the territories of member states, and they will have access at all times to any premises, information and people to the extent necessary to verify the nuclear materials and ensure observance of the regulations specified by the commission. EURATOM was also charged with developing and enforcing regulations for safety and radiological protection.

Each member state had already initiated a nuclear programme at the time the EURATOM charter was discussed. Each had an Atomic Energy Commission or equivalent government agency. Accounting for nuclear materials was obviously an important consideration, as were safety and the control of radioactivity. The members assigned responsibility for formulating these policies, developing regulations and conducting inspections to the EURATOM Commission. Since the member states had the legal and police powers to enforce the regulations and prosecute offenders, they retained the task of enforcing the safeguards and safety regulations. As the importance of physical protection became more obvious, the member states, not EURATOM, developed policies and physical protection measures. In effect, the member states assigned their responsibility for accounting for nuclear materials to EURATOM while retaining responsibility for the enforcement of these regulations and for physical protection of the materials and facilities.

The charter assigns broad rights to the commission and its inspectors but provides no guidance on how the safeguards should operate. The safety and safeguards operations developed gradually as needs appeared. The member states and the nuclear facilities in them were obliged to report to EURATOM, and the facilities were expected to keep track of their materials. The organisation defined accountability regulations, and inspections began in 1959. The regulations were not very strict (also the case in the United States at that time). When the United States, United Kingdom and USSR proposed the NPT Treaty in 1967, EURATOM began to define its accounting requirements in greater detail and to intensify its inspection activities.

The present EURATOM safeguards system is defined in EURATOM Regulation 3227 of 1976. The features of the system that may be of interest to other nations are the objectives and the manner in which they are implemented. The member states had been rivals for centuries. Therefore one objective was to provide mutual assurance that the development of nuclear energy would not threaten any member. The charter authorized the use of nuclear materials for national security purposes and the existence of classified information, but with measures to assure the parties that such activities would not be threatening. A related objective was to ensure that all parties would have equal access to information, resources and technology. While the formal organization is complicated and expensive, many of the measures employed and the knowledge gained by long experience may be of interest to others.

As those present at this meeting will remember, supporters of the NPT and IAEA safeguards were not willing to accept EURATOM safeguards as a substitute for IAEA safeguards. Some considered EURATOM to be a type of self-inspection by closely allied countries. Another important reason was that some other countries that were then negotiating agreements with the IAEA insisted that they be treated in the same manner as Euratom. Especially for the latter reason the EURATOM members agreed to give the IAEA the same authority to negotiate facility attachments, to audit records and to verify measurements independently at the EURATOM facilities as in other states.

OTHER PROPOSED MUTUAL ASSURANCE STEPS

A considerable number of measures have been proposed for bi- and multi-lateral safeguards assurances independent of or in addition to those of the IAEA. The following suggestions might be considered.

A number of countries have announced that their nuclear programmes are solely for peaceful purposes. The degree of assurance that such declarations provide depends on the particular circumstances.

Another activity is to exchange visits of officials and experts to nuclear facilities. Occasional visits of experts may be quite convincing when the nuclear complexes are not too large and the important areas have been made accessible.

National inspectors might be exchanged between countries to perform inspections with their counterparts at the facilities of the respective countries. This form of mutual inspection does not require a formal, multilateral inspectorate with its bureaucracy and costs. There are a variety of arrangements of this nature, which should be quite effective in a number of situations if politically acceptable.

A common complaint about these and more formal multinational inspection arrangements is that proprietary information would be compromised. If proprietary information is a valid issue, there probably are procedural and technical ways to permit the inspection activities needed for assurance while protecting the sensitive information. (The procedures put together by the NPT states that are developing centrifuge enrichment plants to provide IAEA assurance while minimizing the exposure of the technology are described in the appendix to this paper.)

A suggestion discussed during the International Nuclear Fuel Cycle Evaluation (INFCE) was multinational operation of sensitive facilities, such as enrichment, reprocessing and mixed uranium-plutonium oxide fuel fabrication plants.[6] It is not likely that countries would adopt this approach solely for the purpose of providing assurances to each other and to other nations. Joint ownership and operation of such facilities might be economically attractive, however, since such facilities are expensive to design and operate and larger plants may be more cost-effective. While such arrangements require the sharing of proprietary information, as is the case for the tripartite development of centrifuge technology, access could be denied to other possible competitors.

In the case of Argentina and Brazil, relatively simple agreements on means to provide assurances to each other and to other interested nations may be quite convincing. It is also useful to consider that such agreements may set examples for other nations that do not now accept safeguards on all their nuclear materials.

APPENDIX: THE HEXAPARTITE PROJECT

In the early 1970s several countries introduced the gas centrifuge process for enriching uranium commercially on a moderately large scale. The sensitivity of gas centrifuge technology, combined with the flexibility of centrifuge cascade pipe arrangements, presented a difficult safeguards challenge. Indeed, for proprietary reasons centrifuge plant operators had strong reservations about permitting access of IAEA inspectors to cascade areas, yet without that access a fairly quick and easy rearrangement of cascade pipes apparently could permit concealed production of highly enriched uranium.

Dramatic expansion of the industry in the late 1970s added urgency to the shared recognition among technology holders that gas centrifuge plants had given birth to a special and serious safeguards problem that required attention and satisfactory solution. There was a general perception that some degree of access by IAEA inspectors to cascade areas was necessary to look for undeclared materials (feed, product and tails) and altered pipe arrangements and to check on the degree of enrichment in the process area. Nevertheless, some operators hoped that improved perimeter surveillance methods might provide adequate assurance.

A series of meetings between the IAEA, EURATOM and technology holders began in 1980 to discuss, explore and reach consensus on a solution. This effort was called the Hexapartite Project. The participants were the IAEA, EURATOM, Australia, Japan, the United States and Troika (comprised of the Federal Republic of Germany, the Netherlands and the United Kingdom). They sought rapidly to develop safeguards strategies that would be *effective* in meeting the objectives of the inspectorate(s) and *efficient* with regard to the application of resources.

The participants agreed that a safeguards design featuring limited-frequency, unannounced access of inspectors to cascade areas would meet the safeguards objectives and would have three principal advantages over non-access designs:

1. Less intrusiveness into plant operations and lower costs for the operator and the inspectorate(s)
2. Simpler implementation of the model, especially for facilities already operating or under construction
3. Greater availability of pertinent measurement techniques in the near future.

In 1982 the participants reviewed a paper entitled, 'Possible Inspection Activities in Cascade Areas for Limited-Frequency Unannounced Access Applied to Gas Centrifuge Enrichment Plants'. The next step was to test and discuss the limited-frequency unannounced access model at an actual facility. The site chosen was Capenhurst, United Kingdom. In early 1983 the participants reached a consensus on suitable inspection activities.

Recommended safeguards called for both direct visual observation and technical measures inside the cascade areas. Direct visual observation inside a cascade area could reveal changes in piping arrangements, undeclared stores of nuclear material and extra feed and withdrawal stations. One technical measure – the application of seals to agreed-upon pipes and valves or flanges – could reveal tampering with hardware arrangements. It was expected that non-destructive analysis measurements inside the cascade areas could reveal production of highly enriched uranium – if significant quantities were being produced.

The limits on inspection frequency were to be based on production capacities and on the flexibility of the cascades of the individual facilities. To maintain credibility, at least two inspectors would generally appear. Inspector access to a cascade area could be delayed by up to two hours. The duration of an inspection within a cascade area was expected not to exceed about eight hours.

An important objective for inspections within the cascade areas is the technical capability to detect the presence or production of highly enriched uranium. Because of time constraints on inspector visits inside the cascade areas, non-destructive analysis measurements designed for that purpose must be fairly quick – several hours at the most. When applied to achieve confirmation that the enrichment level of uranium hexafluoride gas within cascade pipes is less than 20 per cent, these measurements are called 'go/no go'.

During the last six years two principal non-destructive analysis techniques have been developed for go/no go measurements: the 'deposit correction technique' and the 'two geometry technique'. Both methods depend on measurements of gamma ray emissions that permit differentiation between uranium deposited on the inner walls of pipes and uranium hexafluoride gas in the pipes. Together with particular quantitative assumptions, those gamma ray measurements provide sufficient information to solve equations to obtain estimates of the degree of enrichment of the uranium hexafluoride gas within the pipes.

In general, the existing go/no go techniques have worked well with cascades having large diameter pipes (e.g. 120 mm) but poorly for cascades having small diameter pipes (40–45 mm outer diameter). In some small diameter cascade pipes the uranium hexafluoride gas pressure is low, and the uranium deposits inside the pipes are relatively large. Consequently, unknown or non-constant characteristics of the deposits create large systematic errors that make it very difficult to achieve useful enrichment estimates for uranium hexafluoride gas in the pipes. Go/no go techniques are still under development for small diameter cascade pipes.

The Hexapartite Project illustrates how representatives from several countries can meet and fairly quickly reach a consensus on safeguards measures to be implemented at particular facilities. Experience gained in implementing the technical Hexapartite safeguards measures illustrates that the physical characteristics of plant equipment can significantly influence the technical reliability of some safeguards measurements. In particular, large diameter pipes in the cascade complex of a gas centrifuge enrichment plant facilitate the implementation of reliable safeguards measurements in that area. Other countries considering safeguards for particular facilities might be able to utilize the Hexapartite experience.

Presentation – *Dr Milton M. Hoenig*

VERIFICATION ARRANGEMENTS FOR A BILATERAL NUCLEAR CONFIDENCE-BUILDING REGIME

Among the incentives that may motivate two countries that have not ratified the Nuclear Non-Proliferation Treaty (NPT) to establish a bilateral verification regime for their peaceful nuclear activities are the following:

- mutual confidence-building and assurance of no surprise nuclear developments;

- regional stability;

- assurances that would satisfy nuclear suppliers to engage in civil nuclear trade with the countries.

These incentives may be tempered by a strongly held position that safeguards of any sort on indigenous nuclear activities are unnecessary, as well as by the desire to protect proprietary information and withhold information on the status of civil programmes and even to maintain ambiguity of intentions. The long-term advantages for the countries nevertheless might well lie in working out a bilateral system that builds confidence and is transparent enough to be meaningful, so as to provide assurances that would be at least as effective as full-scope International Atomic Energy Agency (IAEA) safeguards.

While the example of IAEA/NPT safeguards and similar systems may form the essential basis for a bilateral nuclear verification regime, verification arrangements from nuclear and conventional arms control agreements – past, present and planned – provide precedents and examples that also could make an essential contribution.

Multilateral and superpower arms agreements are moving quickly, and by necessity, into an acceptance of intrusive on-site inspections in addition to 'national technical means' (NTM), including satellite observation. Following recent breakthroughs in arms control agreements and negotiations, IAEA inspections are no longer the sole cases of an active on-site inspection regime, nor do they represent the maximum in intrusiveness that is allowed. Greater intrusiveness is part of superpower and multilateral arms agreements in effect and being worked out. Only a few years ago two safeguards experts, in claiming uniqueness for IAEA inspections, observed that '[I]f further nuclear arms control agreements are reached it seems likely that they will rely chiefly on satellite observation.'[7] The opposite now appears to be the case.

Argentina and Brazil should consider examining the desirability of establishing a bilateral nuclear confidence building regime. A special bilateral commission could be established to look into procedures for safeguards and verification. Three important considerations for establishing a bilateral verification regime on peaceful nuclear activities are the intrusiveness of on-site inspections, the use of aircraft or satellites for surveillance, and the desirability and need for third-party certification. These matters are examined here, outside the context of the IAEA/NPT model.

VERIFICATION BY ON-SITE INSPECTION

Some multilateral arms control-related agreements from past years rely on the use of on-site inspection for verification. Under the NPT, for example, regular on-site inspection is part of the IAEA safeguards system to detect and deter diversion of nuclear materials. Other treaties of that era permit even unrestricted and suspect-site inspections, as in the Antarctic Treaty of 1959 and the Treaty of Tlatelolco of 1967, respectively. More recently the IAEA has worked out, through the Hexapartite Working Group, a safeguards system specifically for gas centrifuge enrichment plants that allows 'limited frequency, unannounced access' to the cascade areas of a plant.[8] Yet major bilateral agreements of the past – the Limited Test Ban Treaty, the ABM Treaty, SALT I and the unratified SALT II – rely solely on national technical means (NTM) for treaty verification.

In today's world of increased openness and lessened tensions arms agreements for nuclear and conventional weapons and forces are being negotiated with verification measures, to be applied routinely, that require a degree of intrusiveness beyond the realistic expectations of a decade ago. The bilateral INF Treaty, signed in 1987, the multilateral Stockholm Accord of 1986 and the draft multilateral Chemical Weapons Convention being negotiated in Geneva, for example, call for highly intrusive on-site inspections, combined with unimpeded use of national technical means to verify treaty compliance.

Some highly intrusive verification and inspection regimes are described below. Argentina and Brazil might well claim that examples from current arms control treaties are not applicable to their own situation but rather mainly to the superpowers and European countries or to reductions in nuclear weapons. However, now is perhaps an opportune moment for the two countries to examine successful measures and practices that might be applied to the verification of their own nuclear activities, with a view to perpetuating the current era of great mutual goodwill.

The Antarctic Treaty

The multilateral Antarctic Treaty of 1959 prohibits military activity in Antarctica, including the establishment of bases, military manoeuvres, the testing of any kind of weapon and nuclear explosion, and radioactive waste disposal. To verify this prohibition, designated observers from any of the party states have unrestricted access to all areas of the Antarctic and to all stations, installations and equipment to carry out inspections. The same unrestricted access applies to the use of observation by airplane.

The Treaty of Tlatelolco

The 1967 Treaty of Tlatelolco for establishing a nuclear weapons-free zone in Latin America creates a supervisory regional council, the Agency for the Prohibition of Nuclear Weapons in Latin America (OPANAL), with the power to carry out 'special inspections'. Such an inspection is required when requested by any party to the treaty that suspects that some prohibited activity by another party has occurred or is on the verge of happening. Moreover, a party to the treaty that is suspected or charged with a treaty violation has the right to request and obtain a special inspection to investigate the matter and clear its name. OPANAL must submit its inspection reports to all treaty parties.

The Stockholm Accord

The 1986 Stockholm Document of the Conference on Confidence-and Security-Building Measures and Disarmament in Europe (CDE), which regulates exercises and movements by conventional forces in Europe, marked a watershed because of the scope of its verification measures. The Stockholm Accord, a multilateral agreement between the United States, Canada and thirty-three European nations, is verified by reciprocal on-site inspections and national technical means. The CDE approach to confidence-building is to embed the elements of a bilateral verification regime into a multilateral framework. Nationals in one country inspect installations in a rival country for possible violations, as contrasted with international inspection by a multilateral organization, such as the IAEA, employing international civil servants.[9]

In addition to requiring advance notification and invitation to observers at scheduled ground force exercises of a specified size or greater, the Stockholm Accord permits inspections at suspect sites where there is a question of possible treaty violations involving ground forces. A country that is party to the agreement may request a short-notice, intrusive 'challenge' inspection of activities at a suspect site in another party state, and there is no right of refusal, although there is a quota of three challenge inspections per year with no more than one from the same state.

A challenge by a party to the Stockholm Accord must be answered in twenty-four hours; the inspection must begin after thirty-six hours; and the inspection can last no more than twenty-four hours. The inspection may be carried out from both ground and air (for example, helicopter), and inspectors are allowed to use voice recorders, cameras, maps and binoculars and are to be given 'access, entry and unobstructed survey', except at a limited number of restricted areas referred to as 'sensitive points'. Otherwise inspections are in principle 'anywhere, anytime'. In practice, however, requests have been tempered by a mutual understanding between the parties of what is reasonable and fair. The inspecting state must distribute its report to all participating states.[10] In the view of a former member of the US inspection team the advantages of on-site inspection appear to have outweighed any perceived threat to national security posed by the presence of on-site inspectors.[11]

Chemical Weapons Convention

The forty-nation Chemical Weapons Convention that has been under negotiation in Geneva for the past eight years is expected to be verified by national technical means and by a new international on-site inspection authority, the Chemical Weapons Agency, modelled after the IAEA. The draft convention prohibits the development, production, acquisition, possession, transfer or use of chemical weapons. According to the 1988 'rolling text' of the Geneva negotiation, the Chemical Weapons Agency is to have authority to make short-notice, suspect-site inspections of undeclared facilities if requested by another party. It is to make confidential reports on these inspections to agency officials but not to the requesting country.[12]

Recently a joint US-Soviet recommendation submitted to the Geneva conference created a timetable for the destruction of chemical weapons over a ten-year period. The joint recommendation's proposal for challenge, 'surprise' inspections at suspect sites is very intrusive: monitoring compliance with the Chemical Convention is expected to involve hundreds or even thousands of civil chemical industrial sites around the world.[13] To test the procedures, the two countries signed a memorandum of understanding on 23 September 1989 that calls for exchanges of data and visits to each others' 'relevant military and civilian' chemical weapons sites.[14]

The INF Treaty

The treaty between the United States and the USSR on the Elimination of Their Intermediate-range and Shorter-range Missiles (INF Treaty) went into force in June 1988. An important feature of the treaty is on-site inspection to verify the elimination and banning of ground-launched missiles having ranges between 500 and 5500 km. The treaty eliminates the existing missiles (but not the nuclear warheads) and prohibits the flight-testing and production of the missiles and the production of their launchers.

To monitor compliance the INF Treaty prescribes a highly co-operative and intrusive on-site verification regime. The system relies on the exchange of baseline information, on-site inspections and monitoring, as well as the use of national technical means, that is, satellite surveillance. NTM has been a part of US-Soviet treaties since SALT I in 1972 and continues to play an important role in the INF treaty, which prohibits either party from interfering with the national technical means of the other.

The multi-site inspections under the INF are of four types: baseline (that is, beginning inventory), missile elimination, close-out (that is, closing of a facility) and short-notice check-up. Inspectors of the new US On-Site Inspection Agency (OSIA) and a similar organization in the Soviet Union carry out the inspections. In addition, each side has a permanent team of up to thirty inspectors engaged in continuous portal monitoring at one missile production plant in the other country.[15]

All notifications of intent to conduct inspections and exchanges of data under the INF Treaty are made through the Nuclear Risk Reduction Centers set up in Washington and Moscow in 1987. Receipt must be acknowledged within one hour.

The baseline inspections to check out the beginning inventory of missiles, launchers and facilities against data in the treaty's Memorandum of Understanding were completed during the first two months of the treaty. The elimination and close-out inspections are to be completed in a period of three years; the check-up inspections and continuous portal monitoring continue for thirteen years.[16]

The mission of the On-Site Inspection Agency is to monitor on-site compliance. Key inspection activities consist of counting systems, tracking the missiles to the elimination sites and witnessing their elimination. Information from the inspectors is combined at the policy level outside the Agency with information obtained by national intelligence means and through other intelligence sources to make a verification judgement.[17]

All compliance issues that arise under the INF Treaty are resolved through the Special Verification Commission, a newly created consultative body. The commission can meet on short notice at the request of either party to resolve questions related to compliance and to agree on measures that may be necessary to improve the viability and effectiveness of the treaty. Should either side suspect banned activities by the other, they can address the issue to the Verification Commission.

The treaty permits follow-on, short-notice inspections (by a maximum ten-person team) on a quota basis of twenty per year during the first three years and fewer in subsequent years. Once the inspection team lands at the designated entry point on the other side's territory it chooses a declared site to inspect with a nine-hour warning time. The quota inspections are at declared missile operating bases and support facilities only.

There is no provision in the treaty for challenge inspections ('anywhere, anytime') at undeclared, suspect sites as there is, for example, in the Stockholm Accord. A provision for roving inspectors was apparently discussed during the early INF negotiations but was dropped as being too radical a departure from previous US-Soviet nuclear weapons agreements. Conceivably the Special Verification Commission could be a channel for filing complaints about suspect sites.

In general inspectors are not given free rein. Both sides are expected to take reasonable measures to protect national security information. For example, note-taking by US inspectors during transit to the inspection site is not allowed by their Soviet escorts. There is total coverage by escorts at all times: housing, transportation, inspection activities, meals and leisure activity are closely supervised. The inspections (aside from continuous monitoring at two sites) are limited to twenty-four hours. A written inspection report must be prepared and submitted to the other side in duplicate.

Inspectors are restricted to looking in buildings, vehicles or structures large enough to contain the smallest treaty-limited item and may be denied entry to a structure that is incapable of containing such an item. Site commanders are allowed to shroud non-treaty items that cannot be moved. Inspectors are expected to measure shrouded items to assure that they are too small to contain any treaty limited items – missiles, stages of missiles, launchers and support equipment – and to inspect the shrouded items if the dimensions are greater than allowed.

According to the treaty's inspection protocol an inspection team may bring on to the inspection site tape measures, instant cameras, portable weighing devices, radiation detection devices and other agreed-to equipment.

While there have been situations where a US inspector was denied access to a facility because of a challenge by the Soviet escort on grounds that the facility was too small to hold a treaty-limited item, in most cases the inspectors have been allowed to proceed without interference.[18]

In setting up the INF treaty the two sides agreed to handle implementation and monitoring separately from verification of compliance. They sought an effective system for handling treaty implementation and compliance questions. Under the treaty both sides have OSIAs for implementation and on-site inspection, and they established the Special Verification Commission to resolve compliance questions. Thus four organisations work on implementing the INF Treaty: the Risk Reduction Centers for notification, the OSIAs for inspection, the intelligence community for monitoring by national technical means and other information-gathering methods and the Special Verification Commission for compliance questions. These activities require coordination.[19]

Trends

The trend in arms control agreements is clearly in the direction of intrusive on-site inspections for verification. While this approach is necessitated by the growing complexity of treaty provisions, it also clearly reflects a prevailing mood of confidence-building and cooperation.

While the bilateral INF Treaty falls short of prescribing 'anywhere, anytime' inspections, it does have certain exemplary features that would complement any bilateral verification regime: namely, short-notice inspections at declared sites; communication between treaty parties through a dedicated channel, the Nuclear Risk Reduction Centers located in the respective capitals; and a joint Special Verification Commission to work out any differences that might arise.

The INF Special Verification Commission is similar to but more responsive than the Standing Consultative Commission of the SALT Treaties. It is designed to expedite the handling of problems that may arise. In a bilateral confidence-building regime between countries such as Argentina and Brazil, a Verification Commission could be constituted to resolve problems and complaints, possibly in lieu of specific provisions for challenge, 'surprise' inspections.

An even more basic confidence-building measure would be for the two countries to agree to exchange information regularly on their nuclear facilities. The recent treaty between India and Pakistan pledging no attack on each other's nuclear facilities requires an annual declaration by each party of all such facilities.

The right to challenge inspections at suspect sites is a feature of the Stockholm Accord and the draft Chemical Weapons Convention. Much earlier it was also included in the regional Tlatelolco Treaty in the form of 'surprise inspections' to be carried out by OPANAL, on request.

Challenge inspections at suspect, undeclared facilities are controversial because their implementation could be confrontational. Nevertheless, proposals for such inspections in some form are a feature at the most active negotiations. For example, recently at the Vienna negotiations on conventional force reductions NATO countries introduced a proposal that included the right to request inspections of undeclared weapons facilities suspected of treaty violation but gave suspected violators a 'limited right' to delay or refuse such inspections.[20] More stringent suspect-site inspections without right of refusal have been proposed for the Chemical Weapons Convention and discussed for the START treaty.

Some argue that a suspect-site investigation would uncover illegal activity only if a mistake failed to cover it up and consequently that the right of investigation is of questionable value, even as a deterrent. Others contend that the likelihood of detecting a mistake will increase with improved detection equipment and shorter notice, supplemented by national technical means. One suggested approach would be to allow a right of refusal for suspect-site inspection but with the obligation to provide alternative means to demonstrate compliance.[21] However, the result of such a delay might be only to hide an illegal activity. Alternatively, emergency action by a joint verification commission may attempt to resolve a suspected problem before the short-notice inspection is set to take place.

VERIFICATION BY AERIAL AND SATELLITE SURVEILLANCE

While the verification of peaceful activities at declared nuclear facilities under a bilateral safeguards regime may rely on regular on-site inspections of one party by the other, the identification of undeclared, suspect sites would depend on national intelligence collection by the two countries from a variety of possible sources.

If a quota of challenge inspections is allowed under a bilateral agreement, then the countries involved should have adequate means for gathering intelligence about significant illegal activities either to build confidence that such activities are not occurring or to justify a challenge if such an activity is suspected. The most promising non-intrusive verification measures, borrowing from the national technical means used by the United States and USSR to verify arms control agreements, are photographic-reconnaissance aircraft and satellites.

The proposed use of aircraft overflights for treaty verification and confidence-building dates back to President Dwight D. Eisenhower's 'Open Skies' proposal of 1955. Aerial surveillance can be directly implemented under a bilateral arrangement at a relatively low cost, but it may be politically undesirable. Remote-sensing satellites are by far the least intrusive way to gather photographic intelligence on ground activity. Satellites are the principal means used for photo-reconnaissance by the superpowers today. Commercial satellite imaging of land resources for economic forecasting is now widely available. The best resolution that can be obtained commercially or that is planned is in the 5 to 10 metre range; that degree of resolution could make satellite imaging by commercial satellite a useful adjunct for verification under a bilateral regime, although resolutions of about 1 metre would be preferable.

Countries contemplating the use of satellite imaging for verification under a bilateral arrangement will have to decide on whether the satellite photography is to be gathered and used individually or jointly. There are several possible ways to work out a viable arrangement for using satellite monitoring: the sharing of intelligence by a superpower from its high-resolution military satellite imagery; the purchase of commercial satellite imaging; and the operation of dedicated satellite surveillance systems by the two countries or by a third country that is agreeable to both parties.

The types of clandestine nuclear activities that could be looked for are the preparation of nuclear weapons test sites and the construction of uranium enrichment plants, plutonium reprocessing plants and plutonium production reactors. The chief indicators are the shape and dimensions of buildings, the types of equipment and the supporting infrastructure.

Using Commercial Satellites for Verification

Perhaps the least intrusive means of verification in a bilateral nuclear confidence-building regime is, as noted, the use of imaging from commercial, land remote-sensing satellites. These satellites, which are operated by government-subsidized business organizations, repeatedly scan the surface of the earth from near-polar orbits to provide information for resource planning and other purposes.

Commercial remote-sensing satellites do not have the high resolution of military surveillance satellites. However, photo-imaging from commercial satellites in some situations may be adequate for verification under a bilateral nuclear regime. Civilian satellite imaging is already widely used by the US military, which is the major customer for Landsat and SPOT images.[22] In general, however, available commercial systems are not sufficiently advanced to play a leading role in the verification of a bilateral nuclear non-proliferation regime. Nevertheless, commercial imaging can play an effective supporting role.[23]

Since the 1972 launching of Landsat I commercial imaging satellites have been used to supply information to the public on earth resources and for other purposes. Currently the leading commercial suppliers of remote-sensing are the US Landsat 4 and 5 satellites, which were launched in 1982 and 1984, respectively, and the French SPOT 1 satellite, which was launched in 1986. Both systems' cameras use electro-optical sensors rather than film. Each sensor records the intensity of the impinging light as a number, and the digitized images are stored temporarily on board and periodically transmitted to receiving stations that are located in the line of sight in countries around the world.[24]

Argentina and Brazil each have operating Landsat receiving stations, Brazil has a SPOT receiving station under construction and Argentina has the construction of a SPOT station under negotiation. These stations allow them to receive direct line-of-sight transmission from a satellite.

SPOT provides black-and-white images with 10-metre ground resolution, improved considerably from the 30-metre resolution of Landsat; Soyuzkarta provides 5-metre resolution. Commercial observation satellites from all or some of these sources probably provide limited intelligence data to some countries, such as Australia, Sweden and Brazil, that do not operate their own military systems. Japan is known to have studied Landsat images for information about Soviet activities in Siberia.[25] The SPOT catalogue lists nineteen images of the Pakistani nuclear facilities at Kahuta and fifteen images of the Israeli nuclear facility at Dimona for customers unknown.[26]

Should resolution improve, commercial satellites clearly have the potential for verifying regional nuclear non-proliferation and arms control agreements. However, no launches are planned by SPOT or Landsat with resolution better than the 10 metres currently available from SPOT, even though the French military satellite, Helios, will have a resolution of 1 metre. Another problem, particularly with SPOT, is that regular images of a particular area may not be readily filled because of competing demands.

At the present time available commercial satellite imaging might be used effectively to monitor nuclear proliferation in a bilateral nuclear confidence-building regime in combination with the right of either party to request a challenge inspection of a suspect site.[27] By this means one country would monitor the other's territory for construction activities at possible undeclared nuclear sites or new construction at known nuclear sites. It would then seek clarification of questionable activities through channels established under the bilateral agreement.

Commercial satellite images can show general features. A SPOT image of the Kahuta enrichment plant with its security perimeter in Pakistan shows a massive structure standing out against the adjacent town. The photograph provides evidence corroborating information from other sources on Pakistan's nuclear capabilities. However, the identification of the facility as a centrifuge enrichment plant is not based on features visible in the SPOT photo, nor are there any clues as to whether the plant is configured to produce highly enriched uranium.

Thermal infrared imaging can be used to determine whether undeclared nuclear facilities are operating. The 120-metre resolution infrared images from Landsat 4 and 5 may be valuable for this purpose in some cases. Landsat infrared successfully identified hot spots and the drop in heat discharge to the cooling pond during the 1986 Chernobyl accident. Gaseous diffusion enrichment plants are energy-intensive and are significant heat sources that might stand out on an infra image. However, centrifuge enrichment plants of equal capacity and plutonium reprocessing plants produce less heat.

For nuclear test sites where preparations are being made for testing, satellite imagery would show scarring of the earth's surface from the use of heavy equipment involved in the test preparation and cables connecting the test device in the test shaft to the control centre. It would be difficult and costly to search for unknown test sites, however. Information to verify whether a suspect nuclear reactor is actually operating can be derived from examining infrared images picked up by Landsat 4 and 5 of heat-discharge plumes in adjacent cooling ponds, lakes or streams.

Perhaps the most promising use of commercial satellite imaging in monitoring a bilateral nuclear safeguards regime is the detection of new construction at known nuclear sites. Such information would form the basis for a formal request for further investigation through established procedures in the agreement. While commercial satellite images are able to identify large-scale features of a nuclear facility, important details about nuclear function, capacity or product cannot be determined, except possibly for special features such as round containment domes on large reactors. However, 'before and after' images over a period of months or years can provide the basis for a challenge and a demand for further explanation.[28]

Commercial images are accessible to any party that is prepared to purchase the digitized information from any of the sources and to analyze and interpret it. Neighbouring nations without other intelligence-gathering satellite capabilities may find it effective to monitor a bilateral agreement in this way for purposes of confidence-building and deterrence, even if the overall quality of the information is limited. Observations by public remote satellite could be used to augment intelligence information gathered from other sources.

Thus, despite the limitations in resolution, images obtained from commercial satellites could form the basis for challenge inspections of suspicious sites, and the employment of commercial services may deter parties to a bilateral agreement from pursuing undeclared nuclear development.[29]

Sharing National Technical Means

The United States and the Soviet Union hold a virtual monopoly on the use of earth-orbiting military surveillance satellites. These satellites are one element of the national technical means for intelligence-gathering to verify compliance with arms control treaties and other purposes. Surveillance satellites may carry photo-reconnaissance cameras for imaging in the visible and near-infrared portions of the spectrum, sensors for electronic communications and systems of thermal infrared sensors and radars for the detection of nuclear explosions and for early warning of missile launchings. Photo-reconnaissance satellites are used both for regular gathering of information on military targets and activities and for monitoring crisis areas. Overall, NTM by the superpowers includes surveillance satellites and aircraft, ground-based radars, ground stations for electronic and seismic surveillance, and underwater sensors.

Military photo-imaging satellites are used for two purposes: area surveillance requiring resolutions of 3 to 5 metres and 'close look' requiring resolutions of at least 0.5 to 2 metres. The cost of a complete close-up satellite system has been estimated at $1.5 billion.[30,31] The resolution of current US military photo-reconnaissance satellites for close-up operation is said to be in the range of 10 to 30 centimetres.[32] Resolution depends on camera optics, atmospheric conditions and altitude. Military photo-reconnaissance satellites usually orbit at altitudes ranging between 250 to 500 kilometres. The first French Helios satellite, to be launched in the 1990s, is expected to have a resolution of 1 metre in an orbit of 800 to 900 metres altitude,[33] like the satellites of the SPOT system, on which Helios is based.

A key question is whether either of the superpowers would be willing to provide satellite imaging data on an equal and regular basis to two non-weapons states such as Argentina and Brazil in support of their nuclear confidence-building regime. When it has been in their interests, the United States and the Soviet Union have supplied satellite intelligence information to other countries. The United States has passed on satellite surveillance imagery in quantity to Canada, the United Kingdom and Israel,[34] and some possibly to the Federal Republic of Germany, Iraq and China. The Soviet Union is reported to have provided satellite imagery to Argentina during the Falklands War.[35] Very likely, however, the National Security Council (NSC) and the National Reconnaissance Office (NRO), which operates the satellites and is very protective of its product, would strongly resist any initiative by the US government to share military satellite surveillance data with the parties to a bilateral non-proliferation regime.

Should a superpower choose to assist and encourage a bilateral regime such as Argentina's and Brazil's, it could supply duplicate imaging information about the two countries to both of them. Under one possible arrangement, encoded, digitized electronically recorded imaging information could be transmitted to receiving stations in the two neighbouring countries. The costs to each would involve the receiving station (about US$10 to $15 million) and the data processing and image analysis. More likely the superpower might not opt for direct transmissions for national security reasons. For example, in the early 1980s the United States refused to allow Israel to set up its own ground receiving station for US reconnaissance satellite transmissions. However, the co-operating superpower might be willing to supply already processed and analyzed remote imaging to each of the countries in the bilateral regime.

Using Aircraft Surveillance

There is no question that a mutual 'open skies' policy could do much to assist in building trust between two countries that have entered into a bilateral nuclear confidence-building arrangement as well as to assist in verification. The Eisenhower Open Skies proposal of 1955 was not accepted and then lost importance as satellites took over the major part of the military surveillance effort. Now, in friendlier times, a return to the use of aircraft for surveillance over limited areas would be simpler and cheaper than using dedicated reconnaissance satellites. This approach would be confined to an area of interest rather than covering a wide swath of the earth's surface, unavoidably the case with satellites.

On 12 May 1989 President Bush resurrected the open-skies plan by proposing that unarmed NATO and Warsaw Pact aircraft fly over each other's territory. The Soviets have responded positively to the proposal this time, and an international conference may be called to pursue it further.[36]

Periodic overflights by aircraft could also be used to verify a nuclear confidence-building regime between neighbouring countries. Almost all nations have the capability for aerial photography using manned aircraft or unmanned drones. Information from aerial observation could enhance the effectiveness of the measures for the reciprocal on-site inspection of each other's civil nuclear sites. Periodic airplane overflights for purposes of surveillance, or specific missions to photograph a particular target, would be the most direct and inexpensive means of gathering information to help verify that undeclared nuclear activities were not taking place and to detect any activities that might warrant a request for a challenge on-site inspection.

Reconnaissance aircraft would record information on photographic film with high resolution, comparable to that of military surveillance satellites, and the photographs would be available almost immediately for analysis. Air flights can be launched at short notice and directed at particular sites, while satellites move in predictable orbits and pass overhead at predictable times, a pattern that invites attempts to hide things from their view. For countries such as Argentina and Brazil the cost of aircraft surveillance would be greater than the price of commercial satellite imaging, but the information would be more timely, and the resolution would be ten to one hundred times better.

There are, however, some serious drawbacks to airplane reconnaissance that may make it unattractive to the parties. Overflights would be considerably more intrusive than satellite monitoring. They would have limited coverage, they could be a source of political friction between the countries and they would be subject to attack by air defence forces, an event that would have serious consequences.[37]

Nevertheless, allowing regular aerial reconnaissance flights, and even short-notice flights to designated targets, might be the least objectionable and most feasible regime two countries could agree on if they were serious about establishing an effective system for mutual confidence-building. Aircraft overflights have been adopted as a confidence-building tool under the Stockholm Accord of September 1986.

Commercial or regular military aircraft could be fitted with cameras for photo-reconnaissance missions under an open skies arrangement. Such airplanes, converted to reconnaissance missions, could photograph an area of some 8000 square miles (200,000 square kilometres) per hour from horizon to horizon. Dedicated, high-flying strategic reconnaissance aircraft such as the US SR-71 or U-2, which the United States has long used for military reconnaissance, could also be employed, as well as unmanned remotely piloted vehicles (RPVs), or drones. The SR-71, flying at an altitude of 85,000 feet, can film a ground area of 60,000 square miles per hour.[38]

Airplane reconnaissance can be achieved at substantially less cost than a dedicated satellite. Assuming a country has access to the required satellite technology, the cost of fabricating the satellite depends on the resolution of the cameras: about US$100 million for a satellite with 10-metre resolution, perhaps US$500 million for 1-metre resolution and $1.0 to $1.5 billion for 10-centimetre resolution.[39]

Indigenous or Third-Party Remote Sensing Satellites

Some countries that are potential parties to bilateral nuclear agreements, for example, Argentina and Brazil or India and Pakistan, have space programmes for developing rockets and space launch vehicles.[40] Brazil has been developing the Sonda rocket and Argentina, the Condor. Both countries have plans for developing satellite launch vehicles, and Brazil specifically expects to orbit a land remote-sensing satellite for resource planning in the next few years. There are hopeful opportunities for cooperation in space between the two countries, as evidenced by the signing of the Joint Declaration on Space Cooperation in August 1989, a month and a half after Argentine President Carlos Menem took office.

For the express purpose of verifying a bilateral regime, Argentina and Brazil could operate medium-resolution photo-reconnaissance satellites individually or jointly. Electro-optical sensor technology is available for civil applications, with cost being one of the deciding factors in determining the design resolution of the optical system. For example, both the high-resolution advanced KH-11 (also known as the KH-12) military surveillance satellite, recently launched by the United States, and the civilian Hubble Space Telescope, which is to be launched in 1990, were constructed by Lockheed Missiles and Space Co. and probably have similar optical systems.[41]

To support the aims of the bilateral regime the two countries could also seek assistance from the outside in the areas of launch vehicle, guidance system or optics technology for a joint verification satellite. In return the two countries could announce that they were foregoing the development of a ballistic missile capability for military purposes and agree to adopt the guidelines of the seven-nation Missile Technology Control Regime.[42] Such a co-operative venture in verification could be the outgrowth of the Joint Declaration on Space Cooperation.

For Argentina and Brazil a high-altitude observation satellite with adequate resolution in an orbit that covers the band of latitude between about 57 degrees north and 57 degrees south may satisfy both the needs for verification of peaceful nuclear activities and the scientific and commercial programme of the launching country. The cost of a dedicated observation satellite for verification and peacekeeping is estimated to be over US$400 million, according to recent Swedish studies.[43]

As a further possible means of implementing bilateral verification a third country such as Sweden, Canada or Japan, or a multilateral organization, such as the European Space Agency, could agree to operate a verification satellite and provide satellite surveillance imaging on an equal basis to both countries under a bilateral regime. Both the Swedish and Canadian governments are promoting the development of satellite surveillance for monitoring of regional arms control agreements. Canada has the PAXSAT programme to monitor future multilateral agreements on ground forces and control of arms in space, and Sweden is proposing the Tellus Project for a multilateral consortium to operate a verification satellite with resolution between 1 and a few metres. Japan has had an ongoing space programme, and in 1987 it launched a remote-sensing satellite for ocean surveillance. It plans shortly to launch one to gather earth resource data. By 1993 it will launch another remote sensing earth satellite with a resolution of 8 metres in colour, somewhat better than that available commercially from the French SPOT organization.

Note on Satellite Imaging Technology

Advanced military satellite imaging technology produces ground pictures with high resolution. For example, US close-look reconnaissance satellites are said to be able to 'see' an automobile licence plate from a height of 200 kilometres. The picture of a Soviet nuclear-powered aircraft carrier under construction in a Black Sea shipyard taken by a US military surveillance satellite and published in *Jane's Defence Weekly* shows exceptional detail.

Remote-sensing surveillance satellites carry specially designed cameras to image light reflected from objects on earth in the visible and near infra-red wavelengths. The collected light is imaged on film or on a flat array of solid state electro-optical sensors that digitally record light intensity and may discriminate a particular colour band, although military satellite imaging is usually black and white (panchromatic). Electro-optical devices are also used to pick up thermal infra-red emitted by hot bodies.[44]

The ground image formed by an electro-optical camera is a composition of dot-like picture elements, or pixels, of varying brightness and colour, each formed from the light imaged on a single sensor. SPOT black-and-white pictures, for example, are created by a linear array of 6000 sensors lying in the focal plane of the satellite camera and oriented across the line of flight so as to sweep out a continuous swath on the earth's surface as the satellite moves in orbit.[45]

The more detailed the earth scene that is delineated on a satellite picture, the higher is the resolution. The 'pixel resolution' of a satellite-based camera is defined as the linear size of the ground scene that is 'viewed' by a single pixel at the focal plane of the satellite camera in earth orbit. The size of the ground scene is called the instantaneous field of view (IFOV).[46,47]

Critical to a satellite operation for various applications of remote sensing is the ability to revisit the same area on the earth often. Clouds obstruct imaging as frequently as 75 per cent of the time in tropical or semi-tropical regions and 30 to 50 per cent of the time in temporal zones. Changing events at a time of crisis require frequently repeated views of the same earth scene. Satellites in near-polar, sun-synchronous orbits cover the earth and revisit a place always at the same time of the day. The revisit period of Landsat 4 and 5 (with an altitude of 705 kilometres) is 16 days at the equator. The revisit period of SPOT is 1 to 4 days, less than for Landsat because the cameras can be pointed to either side by up to 27 degrees from the vertical.

Third-Party Certification

Countries involved in a bilateral safeguards and verification agreement would exchange inspectors in order to build mutual confidence. It also may be in their mutual interest to show other nations that the regime is credible, but without sacrificing its confidentiality. A solution to this problem would be to agree on a trusted, third-party certifier to 'look over the shoulder' of the bilateral inspectors by examining inspection records, conducting independent accountancy checks and using its own national technical means to guarantee to the rest of the world, without revealing detailed information, the *bona fides* of the bilateral verification regime.

Why would Argentina and Brazil agree to third-party intrusion in the operation of their national systems, and who would be an acceptable certifier? Having gone so far as to open their civil nuclear systems to each other, going the next step and bringing in a mutually trusted outside guarantor would seem a logical choice. One major benefit would be the opening of supply channels from major nuclear exporters on the basis of assurances of the peaceful intent of their nuclear programmes. For example, the acceptable third-party certification of a bilateral safeguards regime conceivably could, by amendment, be made an alternative to the requirement for full-scope IAEA safeguards as a condition for significant nuclear commerce under the US Nuclear Non-Proliferation Act (NNPA).

The most obvious choice for third-party certifier is another nation. Only a handful of countries probably would pass the tests of acceptability: independent from the superpowers, technically competent, beyond suspicion in their domestic and international nuclear dealings, otherwise acceptable to the two countries, and willing to get involved. Among countries with some of these attributes are Sweden, Switzerland, Japan and Australia. The final choice would be up to the countries that have joined in the bilateral agreement with one consideration being the desired effect on other nations in the world.

Another possibly acceptable choice as third-party certifier would be a private organization that is apolitical and enjoys a worldwide reputation for independence and high integrity. To be viewed as a credible certifier the private company could not be dependent on any national government for a significant share of its income, and it would have available, or be able to acquire, the technical expertise necessary to carry out the required certification activities. For example, a small number of private companies with offices in the major ports throughout the world substantiate and certify the loading onto ships of goods worth millions of dollars. These companies are recognized as final arbiters by shippers and receivers, and their integrity is not questioned.

The drawback to using a private organization as third-party certifier, even a company of the highest trustworthiness, is that it probably would not have the international standing for certification needed to achieve the political aims expected and desired by the countries that have entered into the regional agreement. Yet a private organization, if determined by the two parties to be free of government involvement, surely would not be subject to the same high degree of internal political pressures as an international organization, such as the IAEA. Nevertheless, certification by a nation of recognized neutrality, technical competence and trust in the international community would seem to have the best chance of success.

Finally, certification by a trusted regional organization such as OPANAL, established under the Treaty of Tlatelolco, would be another alternative. The involvement of a regional organization would show that the two countries partaking in the agreement were interested in extending the aura of confidence to their neighbours. However, a regional organization might not have all the technical capabilities required to undertake the certification. In addition, a regional organization, subject to regional pressures, might not gain the degree of international approval that would be required for fully effective certification.

CONCLUSIONS AND RECOMMENDATIONS

The benefits to be derived from a bilateral nuclear confidence-building regime appear greatly to outweigh any apparent disadvantages or inconveniences. The countries involved in such a regime would be able assure each other and the other nations of the world of their peaceful nuclear intentions. The countries would contribute to regional stability, and they would have an impact on the changing world climate of increased openness. Today a range of intrusive verification measures are at the heart of a number of developing and recently completed arms control agreements.

Some of these measures might be readily adapted to a successful bilateral regime.

Argentina and Brazil should consider acting promptly and in their own best interests to examine the desirability of establishing a bilateral nuclear confidence-building regime. A special bilateral commission could be established, possibly under the auspices of the joint Permanent Committee on Nuclear Policy, to look into procedures for safeguards and verification. The verification procedures could include declared peaceful nuclear activities and inspection of undeclared sites. The commission could also look into extending the bilateral verification measures into military activities that are not proscribed by the NPT. The commission should examine the desirability and feasibility of:

- reciprocal, short-notice on-site inspections at each other's nuclear facilities and of inspections, by request of either country, at undeclared sites of the other;

- an open skies policy with the use of observation satellites and aerial overflights for purposes of verification and confidence-building;

- the desirability and feasibility of a mutually acceptable third-party certifier as a means of assuring the benefits from the bilateral regime.

Response – *Dr Miguel Estrada Oyuela*

Although I am an official of the National Atomic Energy Commission, I would like to make clear that my participation here is personal. My opinions do not necessarily reflect the views of the National Atomic Energy Commission.

The members of this panel have presented their thoughts on how to improve 'safeguards'. Safeguards in this context have a very specific meaning, which first appeared in the famous Tripartite Declaration of the three Manhattan Project allies after Hiroshima and Nagasaki. This declaration stated that the dissemination of specialized information relating to nuclear energy before an effective safeguards system was in place would not contribute to a constructive solution to the problem of the atom bomb. It proposed the establishment of an Atomic Energy Commission within the United Nations to regulate the use of nuclear energy.

After that committee was established, the United States put forward a proposal known as the Baruch Plan. In general terms, the plan's idea was that no system of controls based on inspection could be effective. It proposed instead the creation of an international authority that would be the repository of all nuclear material and to which the development of nuclear energy would be entrusted.

The Soviet Union demanded, as a condition for supporting the establishment of this authority, the elimination of all nuclear arsenals. At that time the only nuclear arsenal in existence was that of the United States. Without a doubt a lack of trust and the Cold War climate led to the failure of the Baruch Plan negotiations and to the acquisition of nuclear arms by other countries.

At the end of 1956, UN members approved the by-laws of the International Atomic Energy Agency (IAEA), thereby providing a legal basis for international controls on nuclear material for peaceful purposes.

I do not intend to become involved here in a technical discussion of safeguards. However, I should point out that safeguards are intended to detect, with a high degree of reliability, any diversion of nuclear materials. To ensure a high degree of reliability there must be independent verification systems. In other words, the process of safeguards requires that there be constant suspicion of a diversion of nuclear materials. Safeguards measures must be capable of rapidly detecting any such diversion of a significant quantity of nuclear materials.

IAEA safeguards became the standard within that context of suspicion. Application of these safeguards became the commonly accepted practice and over time the basis for the global non-proliferation regime.

There is still a lack of confidence in international relations. The transformation we are seeing today has not yet produced a new international system.

In the Latin American context, however, the situation is totally different. When Latin America for the first time was threatened by a conflict that was a result of an East-West confrontation – the Cuban missile crisis – its initial reaction was unanimous. We are trying to keep our region free of nuclear weapons. As I see it, this goal has been the success of the Tlatelolco Treaty, which gained the commitment of the region, even among non-parties to the treaty, to remain free of nuclear weapons, which it is. The reason the Treaty has not legally taken effect is that it embodied some elements of the safeguards system that were based on suspicion and a lack of trust. I should, however, be clear on one point – all countries of Latin America have signed the Tlatelolco Treaty and have adhered to it strictly.

The situation in Latin America today is different from that of the countries that had just fought World War II. In Latin America a climate of goodwill exists that itself provides the basis for initiating cooperation. That cooperation produces more confidence, and greater confidence leads to more cooperation. We can anticipate a positive spiral of confidence through cooperation in the peaceful use of nuclear energy for development. Mr Samuel Edlow pointed out some economic disadvantages, but he is looking at it from an accountant's perspective and not from a long-term point of view. Without a long-term perspective we would not be in a position to understand the investments made in the past by today's developed nations. That is why they are now developed, and we are not.

Argentina and Brazil are concrete examples of this confidence between nations. Our programmes complement one another to such an extent that we are able to launch joint development programmes. Under present economic conditions the programmes in both countries will go forward slowly, but they will also do so surely. In the future we will continue to develop confidence in each other.

Let me emphasize just one of these projects, the joint development of fast breeder reactors. This long-term project involves an exchange of key technologies in the nuclear fuel cycle, termed 'sensitive technologies' by those nations that restrict their export. In these restrictions we see the dilemma of non-proliferation. Are we addressing the proliferation of nuclear weapons or the *non*-proliferation of nuclear technology? Argentina and Brazil represent an example of the proliferation of nuclear technology for peaceful uses and, at the same time, the non-proliferation of nuclear weapons.

Response – *Dr Bernardino Coelho Pontes*

I would like to point out that I have taken a one-week leave of absence in order to be free to speak about my previous experience with safeguards. From 1959 to 1961 I used to attend the General Conference of the International Atomic Energy Agency (IAEA) and with some of my colleagues in Brazil I engaged in establishing the very first safeguards system. As a representative of Brazil, I had the privilege of working in the IAEA's safeguards division. I also participated in many discussions to draft safeguards agreements with such countries as South Africa, Israel, India, Portugal and others.

I am astonished this afternoon – I never realized, Mr García Moritán – that our countries were so important and that the danger they present requires something like the establishment of an International Nuclear Central Intelligence Agency. Thus, whenever one of our experts moves about the country, someone will see this fellow who might be constructing a reprocessing plant somewhere, in the Boca suburb in Argentina, for example, in collusion with the Brazilian people. My God, I am really astonished! Nevertheless, this is the beauty of democracy and international affairs.

I took a leave of absence in order to express myself without committing my government to anything I may say here. Now, however, to break the tense atmosphere – because I was tense when I read Dr Milton Hoenig's paper – to break this tension, let me be practical and put my few minutes to practical use.

First, we should recall the conference's scope. This conference is to examine the important aspects of the evolving nuclear cooperation and confidence-building among Latin American countries. The question of this particular panel is, 'What are the available models for the development of a bilateral nuclear confidence building regime?'

I have been presented with two papers, one by my old friend Dr William Higinbotham and another by Dr Hoenig, whom I had the pleasure of meeting at this conference. As far as the first paper is concerned, Dr Higinbotham's skilful introduction provides a concise summary of the whole subject of safeguards. Thus I will limit myself to comments on it.

In his introduction Dr Higinbotham shows the relevance of a national safeguards system to the safety and welfare of the state. I fully agree with him. A national safeguards system is very important because we, the nationals, are much more concerned with one gram of plutonium that might be missing from one of our facilities than are our neighbours or someone in Togo. If a terrorist were to take one gram of plutonium and put it in the water reservoir, it would be a terrible thing. Nevertheless, the IAEA has not given much consideration to national safeguards systems.

I am also pleased with the third part of Dr Higinbotham's paper. I take note of the reference to the EURATOM system. I would like to read what the paper says:

> Next we shall briefly mention some of the features of the IAEA safeguards with which this audience is familiar and describe the EURATOM system in somewhat more detail, not because it may be an example for other states to consider but rather because it may contain some features and represent some experience which might be relevant.

EURATOM safeguards were established before the IAEA, and the IAEA had to convince EURATOM to accept the IAEA safeguards.

Next, Dr Higinbotham made some useful suggestions for which I congratulate him. The first suggestion he mentioned is to exchange visits of officials and to inspect nuclear facilities in both countries, Brazil and Argentina. Indeed, occasional visits of experts could be quite convincing when the nuclear complex is not too large and when important areas have obviously been included.

Well, this system is taking place. As for bilateral inspections, these I cannot assess for the most obvious reasons. We cannot return to prehistoric times. The very first knowledge I had of safeguards implementation was of bilateral safeguards, when the United States and Brazil signed the first nuclear cooperation agreement in this field and immediately implemented a bilateral safeguards arrangement. I think Admiral Thomas Davies must remember. It was the US Government at that time that immediately said, 'We want to transfer to the IAEA the responsibility for safeguards. We do not want the burden of sending our people all around the world. Let us put the IAEA on it.'

Now, we cannot go back to bilateral safeguards. It will be a mess in the world if we have 250 bilateral inspections – in North and South Korea, for example, Cambodia, Vietnam and places like that. The IAEA has the official and approved legitimate power to perform safeguards inspections with its inspectors.

Now I would like to turn to the second topic of this panel and to emphasize that a mutual confidence-building regime between Argentina and Brazil, my colleagues, is not something to be attained. It already exists, not only as stated in written documents but also as proved by the facts. One should remember that not long ago President Raul Alfonsín inaugurated the crucial ARAMAR experimental centre in São Paulo where the first Brazilian enrichment plant is under construction, and some of the units were already in operation. I remember that I flew in in a helicopter with Minister Moritán. He touched our huge centrifuge. Dr Emma Perez-Ferriera was there as well. There is no need to develop confidence-building: these models are already available. I cannot anticipate Minister García Moritán's presentation, but he shows in his paper – and I know that Counsellor José Felicio will also show it in his presentation – that the model exists and could be pursued. Ambassador Julio Carasales already mentioned these points briefly this morning.

As for the model proposed in the second paper, I can only say that such a model is unrealistic and is in conflict with existing legal instruments at

both the international and national levels. Consequently, I will not comment
on it.

Discussion – *Dr Gerardo Quintana*

I want to focus on some concrete steps that relate directly to the establishment of mutual confidence. The kind of mutual confidence I refer to
grows out of personal acquaintance and may be more important than mutual
inspections or commitments obtained from treaties – which often become
empty words.

These personal acquaintances may be an outgrowth of multinational,
professional training courses. Let me describe two courses that have been
offered in Argentina during the past ten years. One is on radiological
protection and nuclear safety, the other on nuclear engineering. In addition
to being useful professionally, these courses have helped create a climate
of confidence among the participating countries.

The University of Buenos Aires and the National Atomic Energy Commission organized these courses, which are funded by the International
Atomic Energy Agency (IAEA). They last eight months. During the
courses, the students carry out their work in Argentine nuclear installations and nuclear plants, in which they obtain a first-hand knowledge of
everything in the country having to do with nuclear material. Over the
past eight years more than four hundred nuclear specialists from Latin
America have taken these courses. More than half the students have been
non-Argentinians. The interaction among professors, who are generally the
principal Argentine nuclear experts, and the Latin American professionals
who take the courses generates friendships that extend beyond the courses
themselves.

These relationships extend beyond the courses over a number of years
by means of professional exchanges and consultations. Many nuclear
professionals have taken these courses early in their professional careers,
and a large number of them now hold important management and decision-
making positions in the various Latin American national nuclear agencies.
Sometimes one can get important information with just a telephone call to
a former classmate.

Such personal acquaintances may do more to create an atmosphere of
confidence than any number of treaties.

I believe the number of courses offered should be increased substantially. They should be offered not only in Argentina but in other Latin

American countries whose nuclear programmes are less well-developed than Argentina's.

The direct personal relationships that result from these professional training courses constitute a warranty of confidence among the countries in the region.

Discussion – *Dr Fernando de Souza Barros*

I will be brief. I would like to make two comments on Dr William Higinbotham's paper. On his first point – the importance of a national safeguards system – he has my complete agreement. He states, 'Any safeguards agreement would almost certainly depend on the national safeguards system of the participating states.' The only thing I do not agree with is his use of 'almost'. I would say any safeguards agreement 'would certainly' depend on the national safeguards system.

The Brazilian Physical Society, in collaboration with the Brazilian Society for the Advancement of Sciences, made this point in a presentation to the House of Representatives of our national congress. Mr Fabio Feldmann, who is here today, witnessed the presentation of our proposal. We asked the national congress to establish a national safeguards system that, in a technical sense, would provide a technical capability for the inspection of Brazilian nuclear projects, installations and facilities that are not under the international safeguards treaty.

I believe that the very basis of any bilateral or multilateral system is essentially a very strong national safeguards system. Building confidence for me implies that when we face any questions with another nation, and that nation has a fairly strong internal safeguards system, any problems that confront us will be minimized by the confidence we have in its national safeguards system.

With respect to bilateral agreements, particularly informal ones, we cannot say they are 100 per cent ideal. When two nations have very friendly relations, any one of these bilateral systems will operate satisfactorily. However, if any issues put these two nations on divergent tracks, we might find that these bilateral systems need something else. I believe this 'something else' is essentially a strong national safeguards system. This is the comment I would like to stress from reading Dr Higinbotham's paper.

To make my contribution short, I would like to emphasize one aspect that Dr Milton Hoenig's presentation called to my attention. He wrote, 'Now is perhaps an opportune moment for the two countries to examine successful measures and practices that might be applied to the verification of their own nuclear activities with a view to perpetuating the current era of great mutual goodwill.' The idea of having bilateral agreements that withstand time and go beyond a particular period of friendly relations seems to me an important point that he brings to our attention.

I do think, however, that the technical contributions he suggests to extend these agreements along these lines might present some difficulties for us Latin Americans, particularly the question of 'open skies' and the utilization of satellites and other modern techniques. I do think that negotiation of an agreement between our nations would serve as a manifestation of a desire on the part of our own people, in this region of the world, to avoid the kinds of confrontations that might lead to the problems we are discussing here.

5 Nuclear Submarines and their Implications for Weapons Proliferation

Presentation – *Dr Marvin M. Miller*

DOES THE ACQUISITION OF NUCLEAR SUBMARINES CONTRIBUTE TO THE PROLIFERATION OF NUCLEAR ARMS?

The implications of the acquisition of nuclear-powered attack submarines (SSNs) for nuclear weapons proliferation is one element in the recent debate about whether various non-nuclear weapons states (NNWS) should acquire SSNs. The general issue can be stated as follows: given their cost, environmental impact and possible connection to the proliferation of nuclear weapons, are SSNs the most appropriate military technology for meeting realistic threats to the national security of particular NNWS? When stated in this manner, the debate about the wisdom of SSN acquisition is strongly reminiscent of the long-standing controversy about the desirability of nuclear power as an energy source in both developed and developing countries, particularly NNWS. In this paper I consider the similarities and differences between nuclear power and SSN acquisition as possible avenues to nuclear weapons proliferation in order to gain insight into the significance of the SSN/weapons linkage. Specifically, I address the following issues:

1. What is the impact on weapons proliferation of the requirement for uranium enrichment and the possible use of spent fuel reprocessing in nuclear power and submarine reactor fuel cycles?
2. Can international and/or bilateral safeguards deter the diversion of nuclear materials from these fuel cycles?
3. Can SSNs and nuclear power facilities serve as political surrogates for nuclear weapons, or, conversely, will they facilitate the acquisition of such weapons?

In the following, I discuss each of these issues in turn and then draw some conclusions about the question posed by the title of the paper.

As many attendees at this meeting will recall, the debate about the connection between nuclear power and the spread of nuclear weapons was particularly spirited during the administration of President Jimmy Carter in the United States. The concern of the 'Carterites' stemmed from the 1974 Indian 'peaceful bomb' and the perception that nuclear power would expand rapidly after the 1973 oil crisis. Although all the official nuclear weapons states produced their weapons materials at facilities dedicated to that purpose, the perceived lesson of the Indian explosion was that establishment of a civilian nuclear power programme could provide a convenient rationale for the acquisition of materials and technologies that were also relevant to the production of nuclear weapons. Particularly worrisome in this regard were highly enriched uranium and plutonium, as well as the 'sensitive' technologies – uranium enrichment and reprocessing – which could produce these materials from natural or low-enriched uranium and irradiated reactor fuel, respectively. A representative view regarding these technologies was that:

> In an important sense, the mere possession of an enrichment or reprocessing plant 'elevates' the possessor to the status of a nuclear weapons state. The fact that bombs *might* rapidly be made leads prudent adversaries and potential adversaries to act as if bombs had been made.[1]

The counter-argument – that international safeguards at enrichment and reprocessing plants would be able to detect and hence deter the production or diversion of weapons-usable materials – was met with scepticism that such safeguards could detect such actions in a timely manner, that is, before their use in weapons. It was also pointed out that safeguards can only be effective when applied, and a state might not agree to safeguards on its indigenous facilities. Alternatively, it could legally withdraw from or violate international arrangements mandating such safeguards.

In sum, the only 'proliferation-resistant' fuel cycles were those in which neither plutonium nor highly enriched uranium was used in separated form. The Carter administration sought to implement this view both by domestic policy and legislation as well as by international agreements. The focus was on measures to minimize the commercial use of plutonium. However, a research programme was also initiated in the United States to develop high density, low-enriched uranium fuels as a substitute for the low density, highly enriched uranium fuels commonly used in research reactors in the United States and abroad. In addition, the United States placed tight restrictions on the export of uranium enrichment as well as reprocessing

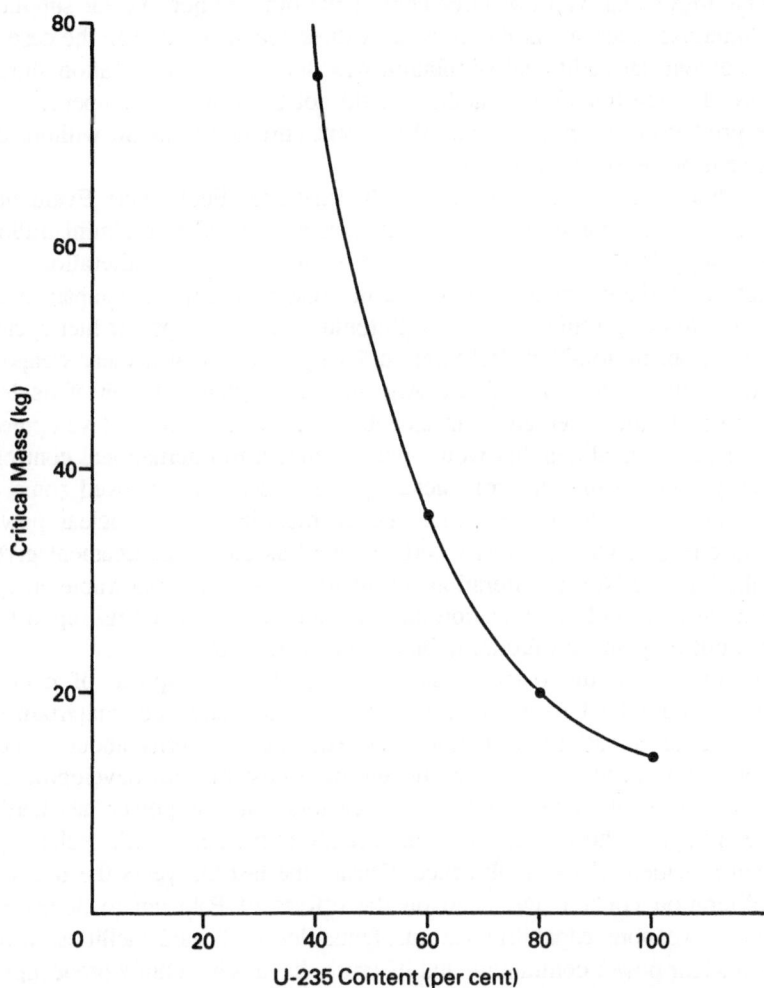

Figure 5.1   Critical mass of uranium as a function of U-235 content[a]

[a] The assumption underlying the design of the device is spherical geometry surrounded by a thick neutron-reflecting uranium shell. Bare sphere critical masses are a factor of three to four times higher depending on the alloy. Imploded critical masses can be significantly smaller depending on the degree of compression.

Source: E. J. Moníz and T. J. Neff, 'Nuclear Power and Nuclear Weapons Proliferation', *Physics Today*, April 1978, pp. 42–51.

technology[2] and won the agreement of the other major nuclear suppliers to 'exercise restraint' in the transfer of these technologies.[3] In the case of enrichment, an additional stipulation was that the recipient nation should agree that the transferred facility would not be designed or operated for the production of greater than 20 per cent enriched uranium without the consent of the supplier nation.[4]

Although the conclusions of the International Fuel Cycle Evaluation (INFCE) programme, organized at the behest of the Carter administration, were supportive of its attempts to raise the level of proliferation consciousness, there was also resistance in Western Europe and Japan to the attempt to delegitimize the use of plutonium in nuclear power fuel cycles. In addition, in non-Non-Proliferation Treaty (NPT) non-nuclear weapons states such as India, Brazil and Argentina the tightening web of nuclear export restrictions served as an incentive for the indigenous development of fuel cycle facilities that would not be subject to international controls. Finally, many supporters of nuclear power world-wide viewed some of the Carter rhetoric, for example, the characterization of nuclear power as an energy source of 'last resort', as well as the encouragement given in the Nuclear Non-Proliferation Act to solar and other renewable energy resources,[5] as a clear indication that the administration and its supporters were not only anti-proliferation but also anti-nuclear.

In retrospect, the fears of the Carterites that the spread of nuclear power would lead to a 'nuclear-armed crowd'[6] have not materialized. Because of a variety of factors, for example, concerns about reactor safety, slow economic growth and the high cost of both developing the required infrastructure and building reactors, nuclear power has hardly spread beyond those states that had already implemented this technology when President Carter took office. During the last ten years the focus of proliferation concern has been on the efforts of Pakistan to develop a nuclear weapons capability via unsafeguarded dedicated facilities, while the nuclear power community and its critics have been mainly preoccupied with the implications of the Three Mile Island and Chernobyl accidents. The hope of many nuclear power advocates is that the development of reactors that have a higher degree of inherent safety than the light water reactor will lead to a nuclear renaissance. In particular the growing awareness of the potential seriousness of the impacts of greenhouse warming – largely the result of fossil fuel combustion – has increased the feeling of optimism in the nuclear community that safer reactors may be the vehicle by which many countries will either revive dormant nuclear programmes or establish new ones.

In my view, the development of such reactors is a necessary but not a sufficient condition for such a revival. An additional requirement is that the associated fuel cycles do not aggravate – rather hopefully ameliorate – the management of radioactive wastes and the concern about the nuclear proliferation/nuclear power linkage. Thus, especially if concern about greenhouse warming increases, we can expect renewed analyses and debate about the proliferation implications of the various safer reactors and their fuel cycles.[7] Meanwhile, the recent plans of several NNWS to acquire SSNs have added a new element to the proliferation debate, to which I now turn.

Historically the development of nuclear reactors for submarine propulsion in the NWS preceded their use as power sources for civilian applications. Indeed, the commercial pressurized light water reactor (PWR) is a direct descendant of the submarine reactor developed for the US Navy under the leadership of Admiral Hyman Rickover in the early 1950s. The obvious difference is that submarine propulsion is a military application of reactor technology, and this fact accounts for the major difference in the implications of SSNs and nuclear power for nuclear proliferation. First, however, I discuss the similarities.

Although most civilian power reactors use enriched uranium as fuel, natural uranium can also be used, as in the Canadian CANDU reactors. However, because of constraints on space in a submarine and the operational requirement for infrequent refuellings, submarine reactors must use enriched uranium.[8] Indeed, current US submarine reactors use uranium enriched to 97.3 per cent U-235, and until recently it was commonly believed that the four other weapons states also used uranium fuels of comparable enrichment for submarine propulsion. However, it is now public knowledge that France developed an alternative low-enriched (less than 10 per cent U-235) submarine reactor fuel technology in the 1970s, and there are also indications that the Soviet Union may be using low-enriched fuel for this purpose.[9]

The proliferation issues raised by the need for enriched uranium fuel for SSNs – and the in-growth of plutonium in the fuel during reactor operation – are similar to those associated with the use of enriched uranium in power and research reactors. In particular, it provides the rationale for the establishment of uranium enrichment facilities under national control, which, in the case of US submarine reactor technology, would be producing uranium that is even more highly enriched than 'weapons-grade' uranium (93.5 per cent U-235). By contrast, the level of enrichment used in the French technology is not sufficient to sustain a fast critical chain reaction (see Figure 5.1). However, even if a technology such as that of the French were adopted, it might be

TABLE 5.1   *Three core designs for a small (50 MW) SSN reactor*

	LEU	HEU	HEU
Power (MW)	50	50	50
Lifetime (years)	10	20	20
Initial U-235 (kg)	74	195	106
Initial enrichment (%)	7	20	97.3
Fuel burn-up (MWD/T)	30,000	60,000	550,000
Final U-235 (kg)	44	124	33
Final enrichment (%)	4.3	14	70
Pu total (kg)	8.9	11.5	0.25
Pu fissile (%)	83	86	76

SOURCE: D. D. Lanning and T. Ippolito, 'Some Technical Aspects of the Use of Low-Enriched vs. High-Enriched Uranium Fuel in Submarine Reactors', presented at the Conference on the Implications of Acquisition of Nuclear-Powered Submarines (SSN) by Non-Nuclear Weapons States, MIT, Cambridge, Mass., 27–28 March 1989, to be published.

LEU   = Low-enriched uranium
HEU   = Highly enriched uranium
MW   = Megawatts
MWD/T = Megawatt days per ton of fuel

relatively easy to reconfigure an enrichment plant whose nominal product was low-enriched uranium to produce weapons-grade material with little loss in separative capacity.[10] The proliferation risks depend on technical factors such as the amount of enriched uranium and plutonium in the reactor cores, the nature of the arrangements that have been made, if any, to safeguard both these nuclear materials and the associated facilities from misuse, and the degree of commitment of the state not to make nuclear weapons.

Regarding the technical factors, details of submarine reactor technology are, in general, closely held, for both proprietary and national security reasons. For example, in the case of US technology nothing is known publicly about the fuel design besides the degree of enrichment. However, on the basis of what is known as well as prior experience in reactor design, a research group at MIT has developed models of submarine reactors with three levels of uranium enrichment: 7 per cent, 20 per cent, and 97.3 per cent.[11] The 7 per cent and 20 per cent enriched designs are based on the French 'caramel' uranium dioxide/zirconium-clad platelet fuel, while the fuel in the 97.3 per cent case is a uranium dioxide/zirconium dispersion cermet. The common assumption is that the reactors are of the pressurized water type with a thermal power of 50 MW. (This level of power is characteristic of a boat of about 2400 ton surfaced displacement with a top speed in excess of 25 knots such as the French *Rubis/Ametheyste*.) The 7 per cent enriched reactor is designed for refuelling every ten years, or every 600 days of full-power reactor operation, while the 20 per cent and 97.3 per cent enriched designs refuel every twenty years or 1200 full-power days.[12]

The results of the MIT study that are relevant to the proliferation issue are summarized in Table 5.1. As expected, in all three cases the amount of plutonium in the reactor core at the end of its life is small compared with that in the fuel discharged from a large power reactor;[13] even in the 7 per cent enriched design there is less than 1 kg of plutonium per year of reactor operation.[14] Current French practice is not to recover the uranium or plutonium in the spent fuel via reprocessing;[15] by contrast, the spent fuel from US submarine reactors is reprocessed to extract the highly enriched uranium, which is then stored or converted into metal for reuse as fuel in the Savannah River weapons production reactors and in the weapons program.[16]

While the uranium in the fresh fuel of the 97.3 per cent enriched submarine reactor design is directly usable in nuclear weapons, 7 per cent enriched uranium cannot be so used. On the other hand, the amount of plutonium produced is much greater in the latter case. The 20 per cent enriched design is intermediate; in particular, while the critical mass is finite, it is much larger than that of 97.3 per cent enriched material (Figure 5.1).

However, as noted, enrichment plants can be converted from the low-enriched to the highly enriched product with a degree of difficulty that depends both on the type of enrichment process employed and on whether the conversion must be carried out so as to avoid detection by outside parties, for example, by IAEA inspectors. From this perspective there may be less proliferation risk where highly enriched uranium in fabricated reactor cores is supplied under stringent safeguards, as compared with a situation in which low-enriched uranium is supplied from an unsafeguarded, indigenous enrichment plant.

Indeed, the question of safeguards was central to the debate about the proliferation implications of SSN acquisition by Canada. Although that debate has now ended with the cancellation of the Canadian SSN acquisition programme, the issue still merits attention, and I consider it briefly below.[17]

According to Article III of the statute of the IAEA, the agency 'shall insure, so far as it is able, that assistance provided by it or at its request or under its supervision or control is not used in such a way as to further *any military purpose'* (emphasis added). This provision implies, for example, that agency safeguards would be designed to insure that enriched uranium supplied for use in a civilian power reactor not be used to make nuclear weapons or in such non-explosive military applications as fuel for naval propulsion reactors or reactors for military reconnaissance satellites.

By contrast, the NPT model agreement, INFCIRC/153, includes a provision (Paragraph 14, 'Non-Application of Safeguards to Nuclear Materials to Be Used in Non-Peaceful Use') that allows a state to withdraw nuclear material in peaceful use from safeguards while it is being used for a 'non-proscribed military activity' such as fuel for a submarine propulsion reactor. As with Article IV of the NPT itself, which guarantees all NNWS parties to the treaty full access to peaceful nuclear technology, Article 14 of INFCIRC/153 was inserted at the behest of several NNWS that were unwilling to forego any of the perceived benefits of nuclear energy beyond nuclear weapons and nuclear explosive devices, specifically including nuclear-powered naval vessels.[18] Thus a NPT NNWS that wishes to acquire enriched uranium for submarine reactors could either invoke the Article 14 exemption or could avoid IAEA safeguards entirely by obtaining unsafeguarded material from a NWS or a non-NPT state. The latter is possible because the NPT only requires safeguards on special fissionable material provided to a NNWS *for peaceful nuclear activities* (emphasis added). Thus both NWS and non-NPT states that have unsafeguarded material in their possession could provide it to a NPT NNWS for non-proscribed military activities without triggering safeguards.

In either case the effect would be that some nuclear material in a NPT NNWS would not be subject to IAEA safeguards. Indeed, this situation might extend to indigenous nuclear materials and facilities as well. For example, a NPT NNWS could build uranium enrichment and fuel fabrication plants for the production of nuclear submarine fuel and claim that such plants need not be subject to IAEA safeguards since they are dedicated to non-explosive military use. However, such a reading of the NPT would do violence to its spirit: there would be no means of verifying that nuclear materials and facilities ostensibly being used in a non-proscribed military activity were not being misused to make nuclear weapons.[19]

For this reason there would probably be strong international pressure on NPT NNWS that wished to acquire SSNs to use the Paragraph 14 mechanism. Indeed, the Canadian government stated that it would invoke Paragraph 14 and also conclude a bilateral safeguards agreement with the submarine supplier. The intent of such an arrangement was to provide continuity of safeguards coverage and hence of confidence that diversion would not occur during the period of exemption from IAEA safeguards, and to do so without revealing sensitive military or proprietary information concerning the design and operation of submarine reactors or allowing international inspectors on board SSNs to verify such information.

The potential problem with such bilateral arrangements from the non-proliferation perspective is that there may be a confluence of interests – commercial, political, strategic – between the supplier and recipient that might tempt the former to minimize any safeguards violations on the latter's part, even assuming that these arrangements were rigorous enough to detect such violations with high probability. This problem might not be serious if the period during which the material was exempt from IAEA safeguards were short compared with the time needed to convert it to weapons-usable form. However, even if the reactor were fuelled with low-enriched uranium, the conversion time would be short compared with the typical ten to twenty year reactor refuelling interval, which is the minimum period during which material would be exempt from international safeguards under Article 14 of INFCIRC/153. Moreover, even after its return to safeguards it would be impossible for the IAEA to use material balance accounting to give *ex post facto* assurance of non-diversion without verified information about the submarine reactor fuel cycle, which, as previously noted, would likely be withheld because of national security and/or commercial considerations.

In sum, even if IAEA safeguards are not bypassed entirely, the burden of proof of the non-diversion of nuclear submarine fuel to nuclear weapons use in NPT NNWS would rest on bilateral arrangements. This point adds to the credibility problem that the international safeguards regime already faces in providing assurance of the non-diversion of significant quantities of nuclear material. Suppliers of SSN nuclear technology and/or materials to non-NPT NNWS would presumably also require some sort of bilateral safeguards agreement. However, indigenously developed SSN-related materials and facilities in non-NPT states, as with their nuclear power counterparts, would not be subject to either international or supplier-recipient bilateral safeguards.

The decision not to become party to the NPT and thus to keep open the option of acquiring nuclear weapons is presumably based on either the perceived military or prestige value of such weapons. The question then arises as to whether SSNs can also fulfil these roles and thus serve as a surrogate for nuclear weapons. On the other hand, will SSNs facilitate the acquisition of nuclear weapons? In this connection, I note that the premise of the Atoms for Peace programme was that most nations would be willing to relinquish the nuclear weapons option if given full access to peaceful nuclear technology. However, for some states the national security and prestige motivations for acquiring a nuclear weapons capability were sufficiently compelling for them to continue along the weapons path. The major achievement of the NPT is that it has delegitimized the open acquisition of nuclear weapons even by states that are not party to the treaty and hence are not legally bound to do so. Thus the taboo against the actual use of nuclear weapons that has built up slowly since Hiroshima and Nagasaki has been reinforced by another taboo on attempts to openly join the nuclear weapons club. One solution to the dilemma of having weapons without violating this taboo – pioneered by Israel and since emulated by other non-NPT states such as India, South Africa and Pakistan – is to indicate by word and deed that the capability to make nuclear weapons exists but actual weapons do not. This phenomenon has been called '*de facto* proliferation'.

Another way for NNWS to demonstrate a mastery of advanced nuclear technology and as a result to acquire a potent military platform while avoiding the onus attached to open acquisition of nuclear weapons is to develop and operate SSNs. Unlike *de facto* proliferation, which is of necessity a 'twilight' phenomenon requiring careful management to make the capability credible without actually crossing the weapons line, there is neither a taboo nor a legal prohibition on the uses of nuclear energy for non-explosive military purposes such as submarine propulsion. Indeed, as noted, the NPT regime provides a specific exemption on the application of safeguards to materials used for such purposes.

For these reasons SSNs may be an attractive surrogate for nuclear weapons that could provide a credible element of 'dissuasion' against excursions of the navies of larger powers, whether friend or foe, into a country's territorial waters and also perform such classic submarine missions as sea denial and coastal defence during wartime.[20] Moreover, the performance of these roles does not require shipboard nuclear weapons;[21] indeed, depending on the military strength of the adversary, particularly with regard to anti-submarine warfare capability, it may not even require SSNs! While it is clear that a diesel submarine, or even diesel-hybrids,[22] cannot do everything an SSN can do, the important issue is whether such boats can do what is needed – in terms of countering realistic threats – at significantly lower costs, taking into account both possible environmental impacts, especially as a result of reactor accidents, and the probable reaction of regional rivals.

This question cannot be answered in generic fashion but only in the context of specific countries. The appropriate solution may be a mixed fleet of SSNs, hybrids and diesels. However, in any case the cost issue must be carefully considered, especially in developing countries where the basic human needs of a significant portion of the population may be unmet. The role of national leaders including the military will be crucial in such decisions, but so is *leadership by example* on the part of the United States and the Soviet Union, where it is increasingly clear that the economic, environmental and social costs of the arms race cannot be sustained. In particular, significant progress in nuclear arms control could create an international climate more favourable to decisions not to acquire nuclear weapons by whatever means: dedicated facilities or the misuse of materials and facilities intended for peaceful or non-explosive military use. With specific regard to the issue of SSN acquisition by NNWS, the following points should be noted:

- There is a widespread consensus among naval strategists that command of the sea in the future lies with the submarine, particularly the SSN, rather than surface ships. This view – reinforced by the acquisition by the superpowers of large numbers of ever-more sophisticated SSNs – provides a strong incentive for the acquisition of SSNs by both militarily significant Third World states such as India, Brazil and Argentina, as well as by members of the western alliance such as Canada, Spain, Italy and Japan.

- The emergence of a new class of 'nuclear submarine states' (NSS) would tend to blur both the psychological and military distinction between NWS and NNWS created by the NPT. As in the case of nuclear weapons proliferation, the degree of opposition to such a development on the part of particular NWS depends on the identity of the NSS. In particular, both the military and non-proliferation establishments in the United States are now opposed to any new NSS, the former because it might limit the freedom of action of the US Navy and the latter because of the perceived risk of increased weapons proliferation. On the other hand, both England and France encouraged the SSN ambitions of Canada but presumably would oppose those of Brazil and Argentina. Finally, the Soviet Union has leased a SSN to India and may also be aiding the indigenous Indian SSN programme, despite strong opposition from the United States.

- Thus a consensus on conditions for SSN supply might not be feasible. However, it may be possible to obtain agreement between suppliers and states wishing to acquire SSNs on criteria that would minimize the attendant proliferation risks. Possibilities include promotion of a low-enriched uranium once-through submarine fuel cycle as an international norm [23] and the development of safeguards arrangements that provide reasonable assurance to the international community that SSN materials and facilities are not being misused for weapons purposes. In this regard a critical issue is whether 'half a safeguards loaf is better than none at all'. That is, can bilateral safeguards arrangements between SSN suppliers and recipients *and* between regional non-NPT SSN aspirants that are developing the technology indigenously, such as Brazil and Argentina, increase confidence in unilateral or NPT non-proliferation commitments? On the contrary, will they serve as a 'marriage of convenience' to blunt international concern about the weapons intentions of SSN NNWS?

- Although SSN-related proliferation risks are real, they should not be exaggerated. As noted, the Carter administration's emphasis on non-proliferation was largely based on the expectation that nuclear power would spread rapidly after the 1973 oil crisis. It did not. For similar reasons, for example, high cost, high and potentially dangerous technology, and stringent conditions on supply, the numbers of states acquiring SSNs will also remain small, at least in the near term. Thus there is time to develop a considered policy toward SSN acquisition.

- To the extent that SSNs serve as a surrogate for nuclear weapons, they may promote international stability: 'better a sub under the sea than a bomb in the basement'. On the other hand, their acquisition might spur naval arms races between regional rivals with no net gain in national or international security. The superpowers cannot hope to minimize this trend by 'advocating water and drinking wine'. Rather, as in the case of nuclear weapons, they should lead by example by decreasing their reliance on SSNs. Indeed, a cogent case can be made that such an action is in their joint interest.[24] US SSNs are the major threat to Soviet nuclear ballistic missile submarines (SSBNs) if they stay under the ice or in coastal waters. On the other hand, Soviet SSNs represent the major threat to US sea control in the North Atlantic, which would be of the highest priority in the event of an east-west land war in Europe. Thus significant mutual reductions in SSNs would address both sides' major concerns and set an important example for other potential NSS.

Response – *Vice Admiral Carlos Castro Madero*

The answer to the question addressed by this panel – does the acquisition of submarines with nuclear propulsion contribute to the proliferation of nuclear arms? – must address three fundamental aspects: the content of the international documents that regulate nuclear activity to avoid proliferation; the possibilities that technological development of propulsion plants offer; and the reaction the acquisition of the submarines can induce in other countries with which there is a possibility of conflict.

If one analyzes the Nuclear Non-Proliferation Treaty, considering it as the international juridical document developed to prevent the proliferation of nuclear arms, the answer to the question is clearly negative. In fact, the objective of the treaty is to avoid acquisition, by any means, of nuclear arms and other explosive devices by countries that did not possess them at the sanction date of the treaty (1968). These countries must sign a safeguard agreement with the International Atomic Energy Agency (IAEA), the conditions of which are established in INFCIRC/153 of the IAEA. This document was written by a committee of government representatives of unlimited composition, called the Committee of Safeguards, and was later approved by the IAEA Board of Governors.

The same document considers the possibility of a country withdrawing from the safeguard system any nuclear material that it may require for nuclear activities not proscribed by the Non-Proliferation Treaty (for example, military nuclear propulsion) and that does not require safeguards. While the procedures for withdrawing such material from safeguards establishes the obligation of the state to require IAEA agreement, it makes clear that it does not mean either asking approval for or providing information on the future use of the material in question.

INFCIRC/153 also establishes that the application of safeguards under other agreements must be suspended as long as there is still an agreement in force of the type of the above-mentioned document, with the sole exception of agreements related to IAEA projects. That is, nuclear material previously perpetually subject to safeguards that prevent its application to military use may be used in military propulsion.

Also relevant to this question is the opinion of the Secretariat of the IAEA in response to a request by the Argentine representative on the Board of Governors as a result of the presence of British submarines with nuclear propulsion in the South Atlantic conflict (Report GOV/INF/433 of the IAEA). Argentina asked for a study to determine the degree of compatibility between the safeguards agreement in force and the IAEA statute that refers to the statutory legitimacy of military applications of non-explosive nuclear materials subject to the system of safeguards. The objective of the IAEA, in contrast to the Non-Proliferation Treaty, is that nuclear energy should not be used for *military purposes*. The report established that the differences between the various types of agreement do not convey any incompatibility between them and the statute. In other words nuclear propulsion is not incompatible with a nuclear programme exclusively directed to peaceful ends.

It can then be concluded that the international community understands that military nuclear propulsion does not contribute to the proliferation of nuclear arms. Canada is a very concrete example of this point. Throughout all its governments Canada has assumed a position of leadership on the subject of non-proliferation. It has always aligned itself rigidly with all the initiatives on that subject. In June 1988 the government of that country announced it was going to incorporate into its navy submarines with nuclear propulsion to control and patrol the coastal waters of the three oceans that border its territory.

Let me now analyze the initial question from the technical point of view, that is, how does the development of nuclear propulsion systems contribute to the acquisition of a nuclear bomb. Let me first point out some fundamental concepts. The proliferation of nuclear arms is an eminently political and non-technical subject. All countries that possess nuclear arms today obtained them through programmes specifically directed to that purpose. Consequently they have followed the shortest and most economic route toward the objective pursued.

There is no question that any development of applications of nuclear fission enhances a country's potential capacity to produce nuclear explosives. However, the decision to make a bomb is a political one. Following that decision all the necessary measures must be carried out. If a country has all the required infrastructure but no political decision, acquisition of a bomb is put aside. In short, what counts is the political intent.

It is unimaginable that a country that may want to develop a nuclear explosive would choose an indirect route such as the development of a nuclear propulsion plant. A propulsion reactor is a compact reactor of enriched uranium. Enrichment of the nucleus is not as high as that required to produce a bomb, nor is this kind of reactor suitable for producing plutonium. Therefore, having a reactor for propulsion is the same, for purposes of this analysis, as having a new variety of the many that are operating in the world without anyone claiming that they represent a possible violation of the status quo.

Finally, let me analyze if the acquisition of a submarine with nuclear propulsion could induce a neighbouring country, with which there exists a situation of potential conflict, to produce a nuclear arm. However theoretical this analysis may be, I can point out the following. The imagined reaction would be totally out of proportion. Nuclear propulsion is a component of a conventional system of arms. One could therefore conclude that a similar reaction would be produced by the incorporation of any system of totally non-nuclear arms that would alter the pre-existing balance of power. Furthermore, the country that reacts would have to have a significant scientific-technical infrastructure in the nuclear area that would probably be subject to IAEA safeguards. It would therefore be more viable to react by producing submarines with nuclear propulsion so as not to confront the serious consequences arising from a violation of the international commitments that everybody respects.

To conclude and to answer the original question of this analysis, in my opinion submarines with nuclear propulsion do not contribute to the proliferation of nuclear arms.

Discussion – *Dr José Goldemberg*

I will divide my comments into two parts. First, I would like to summarize a very interesting article that was written and published recently by Admiral Mario Cesar Flores, Chief of Staff of Brazil's Navy and the leading figure in Brazil's nuclear submarine programme. The article is entitled 'Nuclear Powered Submarines'. It was published in a Navy magazine in 1988.[25] A very interesting article, it attempts to answer the kinds of questions Admiral Carlos Castro Madero just posed. In a way the substance of the article is quite similar to the remarks made by Admiral Castro Madero.

First, Admiral Flores asks the following question: why submarines in the Brazilian Navy? The reason he gives has to do with the Malvinas War. Submarines played a very important role in the war's outcome, Admiral Flores believes – not only nuclear submarines, but submarines in general.

Second, he asks the question: why have nuclear submarines? He gives three reasons why nuclear submarines are preferred. First is their ability to operate undetected. Second is the great distances they can navigate and their velocity, which is greater than that of conventional submarines. Third is the length of time they can remain underwater. Flores concludes that for these reasons a nuclear submarine is more attractive in some conditions than other kinds of submarines.

Flores then addresses the strategic question, which I think is the most important point in his article. He makes statements that have a lot to do with what we discussed yesterday and today.

First, the basic premise of his paper is that Brazil must have a military power adequate to meet its security needs in the world today. In the long term, he writes, Brazil's increasing international responsibilities and interests will require Brazil's greater presence in the distant waters of the South Atlantic. Thus nuclear propulsion must be encouraged. Flores argues that if the problem was only how to protect Brazil's coasts, conventional submarines could do the job. However, as the importance of Brazil in the international scene increases, Brazil's military presence farther away from its coasts will become more important. These are the reasons, then, for developing an indigenous Brazilian submarine, and a nuclear submarine programme in particular.

Secrecy was fundamental to the development of such a project because it would have encountered serious opposition had the public known right from the start. Secrecy is no longer needed except to protect industrial secrets.

Flores goes on to explain that the autonomous nuclear energy project was started because Brazil clearly recognized that the agreement with the Federal Republic of Germany (FRG) would not give it the capabilities it needed to build nuclear submarines. It would, for example, be very difficult to obtain an enrichment capability under the agreement with the FRG because that technology was subject to safeguards.

Brazil's autonomous programme is completely without safeguards. Initiated in 1980, the programme is long term. A by-product of the autonomous programme is that Brazil will have the capability to construct small nuclear reactors, of about 50 MW, and eventually to construct industrial reactors to produce electricity.

The autonomous programme is going very well. There is a uranium enrichment plant in the State of São Paulo that uses ultracentrifuges developed with local technology. Brazilians take great pride in this accomplishment, which was thought to be impossible a few years ago.

Flores concludes by stating that it may indeed be asked if this project is really a cover for the production of atomic weapons. He answers in almost the same way Marvin Miller did in his paper. Flores writes that, although production is technically feasible, Brazil made a 'national decision', which is also the wording used by Admiral Castro Madero regarding Argentina's programme, that Brazil will use nuclear energy only for peaceful purposes. Flores considers nuclear-powered submarines to be a peaceful military activity and not a nuclear-explosive type of activity.

Admiral Flores concludes with the observation that under the terms of Brazil's new constitution, the national congress will have to approve projects of this kind.

Flores' article is very interesting, and I highly recommend that people interested in the subject scrutinize it carefully. I think it represents very well the current thinking of the Brazilian Navy on these matters.

Now, let me add my own views on this subject. First, there is indeed a sense of pride in Brazil that results from these considerable achievements. Second, this project is a very, very long-term one. Just the other day Admiral Flores told me that no nuclear submarine will be operating in Brazil for another twenty to thirty years. There is still a lot to be learned about conventional submarines.

This delay is not a matter of virtue but of necessity. Funding is a very serious problem. Countries such as ours have many social needs, and people are quite sensitive to this consideration. With the return of democracy to our countries it is very difficult to suppress the demands of segments of the population. Therefore, the Congress is scrutinizing budgets in Brazil in a way that never happened before.

Military expenses are also going to be scrutinized. This point holds true for the nuclear submarine programme, which clearly is expensive. Brazilian scientists have taken the position, as I have taken in the Superior Council for Nuclear Energy that advises the President, that the development of science and technology in Brazil embodies many, many needs. Uranium enrichment and the development and construction of nuclear reactors are not to be excluded but must be weighed against Brazil's other development needs. It is not a programme that should have absolute priority. Rather, it must take place within the framework of priorities established by our country's civilian authorities and by Brazilian society.

Finally, the programme has to be placed under very strict internal safeguards. There may be legal questions as to the distinction between international safeguards and internal safeguards. We understand internal safeguards to be a system by which civilians – the civilian government, civilian representatives in the parliament and civilians out of parliament – know what is going on. The knowledge of what goes on is a method of safeguarding against adventures of one type or another. I also think the establishment of a very strong internal safeguards system is a preliminary step to the establishment of a bilateral safeguards system between Brazil and Argentina, as well as to an international safeguards system that could be worked out later.

6 Linkages Between Vertical and Horizontal Non-Proliferation

Presentation – *Martin Gomez Bustillo*

GENERAL INTRODUCTION

When nuclear weapons appeared on earth, they upset the foundations on which international relations had evolved. The concepts of security, power balance, neutrality in conflicts and many others acquired new meanings and in some cases lost their old ones. The capacity of nuclear weapons for destruction and their endless perfecting has introduced the prospect of a holocaust. While there are only a few nuclear weapon states, they somehow have become the arbiters of humanity's future.

The international community, which is justifiably and deeply worried about this situation, has been trying to find ways of avoiding proliferation and to banish the risks of nuclear war. It is known that disarmament is a complicated issue and that the balance between security and disarmament has never been simple or mathematical. However, at the very least there has always been a will to negotiate. Excuses such as 'this is not the right time', 'the time is not yet ripe' or 'is too complex', or 'international circumstances are not propitious for starting negotiations on such transcendental subjects' are unacceptable. Negotiations may be slow and difficult, but they have to be started. By doing so each country makes evident the sincerity of its aims. The international community cannot expect these negotiations to be successful right away, but it has the right to demand that they take place and are not postponed forever, as we are conscious of being involved in a race against time.

PROLIFERATION

In anticipation of gaining greater understanding of the commanding need to move non-proliferation ahead and to halt proliferation – the latter in

171

its broadest sense – it is necessary to start by asking what exactly has proliferated and what are the risks connected with that proliferation for humanity as a whole?

To summarize, it can be said that since the first nuclear weapon test in July 1945 in the United States, a testing that continued with the inhuman utilization of the bomb on the battlefield – first in Hiroshima and then in Nagasaki on 6 and 9 August 1945, respectively – and up until the latest tests carried out by the nuclear weapons states, which have varied in their number and destructive capacity, nuclear weapons have proliferated to such an extent that the number of warheads in the world has been estimated at 50,000. The total capacity or explosive potential of existing nuclear weapons is approximately equivalent to one million bombs such as the one dropped on Hiroshima, which had a power of 13 kilotons (remember that 1 kiloton is equivalent to 1000 tons of TNT).

According to recent studies by the United Nations on nuclear weapons, it has been estimated that should a nuclear exchange of 10 000 megatons occur in the event of a nuclear war (1 megaton equals 1 million tons of TNT) and only half that quantity were dropped on cities, more than 1000 million people would die instantly. Added to this result are the climate changes that use of these weapons might produce, even in regions far from the ones attacked. Those changes would include, among others, obstruction of sunlight for several months, abnormal temperature drops and a reduction of the protective ozone layer in the stratosphere that would expose the earth to ultraviolet rays.

Unfortunately, mankind has not created just this type of mass destruction weapon based on fission and fusion. It has also developed chemical weapons. The latter were responsible for approximately 1.3 million dead and injured during World War I, a source of inexplicable human suffering and devastating impact on the lands of Belgium and France. During the Iran-Iraq War their usage also caused thousands of dead and injured, despite very strong condemnation of their utilization by the international community both from a humanitarian viewpoint and as a flagrant violation of the Geneva Protocol of 1925.

These, among others, are the reasons for which it is important that both nuclear weapons and non-nuclear weapons states become aware of the danger of their use and carry out concrete international negotiations to end this type of proliferation, reach an agreement on their prohibition and eventually dispose of these weapons of mass destruction.

Conceptual Definitions

I have referred to the concept of 'proliferation' several times. In its broadest sense this concept may be defined as 'the multiplication of similar forms', that is, as the production of further units of something. According to Webster's *Collegiate Dictionary*, 'to proliferate' means – in its transitive form and textually – 'to grow by rapid production of new parts, cells, buds or offspring; to increase in number as if by proliferating; to multiply'. As an intransitive verb it is 'to cause to grow by proliferation'.

The above concept, when used within the specific context of disarmament, refers especially to the diffusion or multiplication of mass destruction weapons. However, when dealing with the proliferation of weapons, a distinction must be made between the various categories of weapons, since, depending on their features, some may have large-scale lethal effects comparable to those described earlier.

As a general rule, when reference is made to the proliferation of weapons, the reference is to weapons of mass destruction such as nuclear, biological and chemical. However, the so-called conventional weapons, with their own characteristics, should not be discounted with respect to their lethal effects or their high economic cost.

Nuclear weapons, however, as a category by themselves, have a peculiar feature: they may release – within less than a second – a large amount of energy, which is generated by the fission and fusion of the nuclei of atoms. The resulting explosion – involving a blast wave, heat and fire plus immediate and delayed radiation – causes tremendous devastation. As an illustration of their power, a single nuclear weapon can release an amount of energy greater than that released by conventional weapons in all the wars of history.

When the expression 'disarmament' is used, it refers to the control of weapons, to their limitation, to non-armament and to measures for the creation of confidence, the goal being to reduce and then eliminate weapons.

This theoretical definition of 'non-proliferation' becomes more complex when applied to avoidance of the multiplication of the knowledge, materials and equipment needed to fabricate nuclear and other types of weapons. That is, the objective of non-proliferation becomes more encompassing, for example, extending to the proliferation of knowledge on the development of launching systems, the fabrication of vectors, remote control steering and highly complex computers, given that such knowledge is a basis for the integral development of nuclear weapons.

To summarize, non-proliferation of weapons not only includes a prohibition of the multiplication of war devices, but it also extends to the means and elements necessary to produce them. Such measures are aimed at avoiding the proliferation of 'nuclear weapons states'.

When reference is made to 'nuclear weapons states' and to 'non-nuclear weapons states', NWS and NNWS, respectively, in the field of disarmament negotiations, the aim is to note and distinguish explicitly the difference between countries 'having nuclear weapons' in the first case and 'not having nuclear weapons' in the second one. A clarification must be made regarding this classification, since some countries have attained a very high level of development and knowledge in nuclear technology yet do not have nuclear weapons, a praiseworthy characteristic. These countries are considered NWS in the context of science and technology and as NNWS for purposes of disarmament and non-proliferation.

Types of Proliferation

Ever since people stated the need to start negotiations and reach agreements on disarmament, they have used the expression 'non-proliferation' repeatedly, it being considered as a central issue.

This concept, as defined above, embodies three types of classes of proliferation as follows:

1. Vertical proliferation, which refers to the deployment of nuclear weapons by the five NWS. That is, it produces an increase in the sophistication, number and destructive capacity of the nuclear weapons available in NWS.
2. Geographic proliferation, which involves the deployment of nuclear weapons by the five NWS outside their national territories, either in countries under their dependence, in still other countries or in the air and sea space outside their national jurisdictions. That is, this proliferation occurs when nuclear weapons are positioned not only in the countries that produce them, but over a more extended geographic area, such as in territories belonging to allies of the NWS. The presence of submarines and surface boats carrying mass destruction weapons is typical of this type of proliferation, and extends the possibility of destruction to all seas and oceans in the world.
3. Horizontal proliferation, which occurs when additional countries acquire nuclear weapons – either through transfers or their own production. While the other types of proliferation are undeniable,

this type differs in that the weapons are presumed to exist, without actual confirmation.

THE TREATY ON THE NON-PROLIFERATION OF NUCLEAR WEAPONS

In the context of the above discussion, it is clear that proliferation affects the whole international community and is not a particular or exclusive preoccupation of the NWS or other countries that have signed the Non-Proliferation Treaty (NPT). The first countries signed it on 1 July 1968 in the cities of London, Paris and Moscow. It came into force on 5 March 1970. By late 1987 137 countries had become signatories, including three NWS: the Union of Soviet Socialist Republics, the United States and the United Kingdom. Although not signatory, France has indicated it will respect the treaty's content. China, on the other hand, has repeatedly criticized the treaty, stating that it is discriminatory. China has also pointed out repeatedly that it does not support nor does it promote nuclear proliferation and that it is not aiding other countries to produce nuclear weapons.

Since the treaty took effect, several review conferences have been held (1975, 1980 and 1985), and another is to be held in 1990.

The treaty contains eleven articles. Their analysis reveals three fundamental obligations for the countries involved:

1. NNWS cannot acquire such weapons or nuclear explosives (Article 2). Further, they must submit all their nuclear material and activities, available internationally or developed on their own, to the control (safeguards) of the International Atomic Energy Agency (IAEA), headquartered in Vienna (Article 3).
2. As per Article 4, nuclear developed countries guarantee that they will facilitate the exchange of equipment, materials and technology and that they will not hinder the development of production and peaceful uses of nuclear energy by all the treaty's member states.
3. In Article 6 the NWS promise to end the nuclear arms race and to negotiate a general disarmament treaty in good faith.

It is also important to remember that under Article 5 the parties promise to adopt measures to ensure, under international observation, that the potential benefits from any peaceful application of nuclear explosions will be made available on a non-discriminatory basis to the NNWS that are members of the treaty and that those countries will not be charged for related research and development.

Analysis of these articles leads to a series of reflections. First, the treaty refers to two very different types of participants – the NWS and the NNWS. Article IX clarifies which are the NWS. The third paragraph, second part, reads: 'For the purposes of this Treaty, a nuclear-weapon State is one which has manufactured and exploded a nuclear weapon or other nuclear explosive device prior to 1 January 1967.' This arbitrary date was aimed at freezing this group of states.

Obligations Concerning Nuclear Weapons

The obligations and rights established for each category with respect to nuclear weapons and the peaceful use of nuclear power are very different. Below is a description of their content.

Paradoxically, most obligations regarding nuclear weapons fall on the NNWS. To guarantee they carry out their obligations as parties to the treaty, the NNWS are supposed to open their facilities to a system of inspections and safeguards and to submit their technology, material and equipment transfers to the IAEA's control.

For their part the NWS are only required to enter into negotiations aimed at ending the nuclear race and beginning disarmament on a date close to 1968. These countries did not and have not assumed any obligation concerning the opening of their facilities – either peaceful or defence-related – to international control. Nor are they obliged to avoid using nuclear weapons against the parties of the treaty or prevented from threatening them with their use.

Even more different than the two groups' obligations concerning warlike matters is the way in which those obligations are being implemented. Ironically, not only did the NWS not discontinue the arms race in this field, they accelerated it. The numbers of weapons multiplied enormously and became more sophisticated. Nowadays, arsenals are ten times as great as they were fifteen years ago, while their effectiveness has increased prodigiously. The NWS are also carrying out frequent nuclear tests for various civil and military purposes.

In contrast, the NNWS have adhered strictly to their promise. Even those countries such as Argentina that did not sign the NPT have abstained from manufacturing atomic bombs. Only India has detonated a nuclear bomb (once in 1974), while stating that the technique was developed for peaceful purposes.

To summarize, it has been stated that the failure of the NNWS to develop nuclear weapons is proof of the treaty's effectiveness. However, the countries that have not signed the treaty have also not developed nuclear weapons. Meanwhile the danger of a nuclear confrontation has grown tremendously as a result of the arms race in which the NWS are involved, countries that talk about 'limited nuclear war' and 'clean' or 'surgical nuclear weapons' that would not result in generalized destruction of the planet, but instead limit their effect to specific areas.

Obligations Concerning the Peaceful Use of Nuclear Energy

The issue of nuclear explosions for peaceful purposes is one of the obstacles preventing some countries such as Argentina and Brazil that want to keep their right to profit from the benefits they might obtain from such technological development from signing the NPT. It is worth remembering that the Tlatelolco Treaty, signed one year before the NPT, allows member parties to perform tests for such a purpose. Moreover, to compensate for the restrictions imposed on weapons the NPT promised to facilitate the access of the NNWS to technology and equipment that could stimulate their nuclear development for peaceful purposes. However, this access has been very relative. Except in Europe, Canada and Japan, where technological and industrial development were promoted for political and economic reasons related to their involvement in the western alliance, not many countries have achieved significant progress in the peaceful use of this type of energy. Among them are Argentina and India. It must be asked whether their progress was possible despite not having signed the NPT or because of that decision.

This question is not pointless considering that even though the treaty specifies an obligation to enter into technology transfers at the same time such transfers are subject to unilateral restrictions by the NWS. More precisely, the US Nuclear Non-Proliferation Act of 1978 states that 'the proliferation of nuclear explosives or the capacity to manufacture or somehow obtain them poses a serious threat upon the US security, as well as upon the progress of peace and development in the world'. In short, a country's unwillingness to manufacture atomic bombs is not enough. It cannot even have the capability to manufacture them, no matter that that capability is aimed exclusively at peaceful use. Undoubtedly, this unilateral limitation on technology transfers is liable to be applied based on political and commercial criteria. In fact, the act states that a purpose is to discourage peaceful nuclear development: it is the United States' policy

> . . . to cooperate with foreign nations in the identification and adaptation of technologies that are adequate for the production of energy and, particularly, to identify alternative options to nuclear energy, in order

to consistently make their energy needs compatible with their material and economical resources and with environmental protection.

This goal is praiseworthy, and I share it. However, up to now those options are neither feasible considering the future exhaustion of traditional sources of energy nor effective when they are to be applied on an economic scale.

Control Over the Peaceful Use of Nuclear Energy: Safeguards

The control and safeguards system to which the NNWS are subject have no equivalent with respect to the NWS, although the latter are the most advanced in applying nuclear energy for peaceful purposes. That application is an industrial activity involving huge investments aimed at the development of a product with high value added. Its marketing is extremely interesting, one reason for the increase in competition in the international technological market. It is therefore surprising that NPT developing countries in this branch of business are obliged to accept a control regime while countries that are more developed in this field reserve their rights not to behave reciprocally in this respect.

This situation reflects flagrant discrimination, with some countries having to run the risk of violating industrial secrecy, while others are not obliged to do so. Not too much needs to be said about the damage that such a violation may cause the marketing of technology and equipment projects involving the peaceful use of nuclear energy.

NUCLEAR NON-PROLIFERATION UNDER THE NPT

Efforts Toward Vertical and Horizontal Non-Proliferation Under the NPT

Concerning horizontal proliferation
Article I of the NPT establishes that the NWS which are parties to the treaty undertake not to supply nuclear weapons nor to assist in their manufacture by the NNWS.

According to Article II, NNWS that are party to the treaty agree not to receive or manufacture nuclear weapons.

Article III establishes that, to prevent the diversion of nuclear materials for military purposes, NNWS parties must subject their peaceful-use facilities to international safeguards.

The goal of ending proliferation does not invalidate nuclear cooperation for peaceful purposes, as shown in Articles IV and V, in which the NWS promise to aid NNWS parties to the treaty in their peaceful use of nuclear energy.

Concerning vertical proliferation

Article VI establishes that the NWS are responsible for starting serious negotiations aimed at ending the arms race. The NWS are far from meeting that promise, given that their nuclear arsenals have extended beyond the limits of the imagination. Moreover, the sophistication of war devices progresses unceasingly. Technological progress, which could provide many benefits to humanity, is being promoted and used for profit to increase destructive capacity. The time available for moving non-proliferation from theory to reality is becoming shorter and shorter.

In addition disarmament negotiations, which also are not progressing as they should, have amounted to a 'step backward' in the hope for life. In spite of the oft-repeated claim that 'a nuclear war cannot be fought or won', any inactivity in this field contrasts with the qualitative and quantitative arms race that is growing day by day. It must be questioned whether the conclusion of the INF treaty and the ABM agreements are the start of carrying out Article VI to end the nuclear arms race. They may be a beginning, but a lot more has yet to be done.

Concerning nuclear weapons tests, if the NWS discontinued them and their race to improve their nuclear weapons, they would have no reason to continue with new tests, considering that their first objective is updating nuclear weapons and making them more effective. The NPT explicitly calls for negotiations aimed at achieving a comprehensive test ban treaty. Again, the world is far from that goal, given that the superpowers instead of discontinuing their tests are carrying out joint experiments in the verification of nuclear testing.

Should the five NWS fully meet their promise and responsibilities regarding proliferation and the discontinuance of nuclear weapon tests, a new and promising political situation would arise. A comprehensive treaty banning nuclear weapon tests would enormously strengthen the efforts toward preventing the proliferation of nuclear weapons.

Characteristics of the Non-Proliferation Efforts

On the basis of the content of the previous paragraphs, I believe there is discrimination or asymmetry in the efforts toward vertical and horizontal non-proliferation and that the positions of the NWS and NNWS are unequal: Articles I and II establish effective obligations (verifiable in the case of the second), while Article VI only establishes an obligation to enter into negotiations.

A question may be raised in connection with the implementation of Article IV, second paragraph, of the NPT, where all the parties promise to 'facilitate' the fullest possible exchange of equipment, materials and scientific and technological information and the transfer and assistance concerning knowledge on the use of nuclear energy for peaceful purposes.

Within that framework of international cooperation it must be noted that, in spite of the obligation spelled out in Article IV, no assistance has been provided. No nuclear power or desalinization reactors have been given or sold to any country as a direct result of that article.

The greatest contribution the NWS could make toward cooperation on the non-proliferation of nuclear weapons or compliance with the philosophy of the NPT would be immediate negotiations and signing of a threshold agreement or a total prohibition of nuclear weapons tests. These agreements would end the continuing evolution of more and more sophisticated weapons (vertical proliferation) and the expansion of the nuclear arms race.

Efficacy of Non-Proliferation

Has vertical non-proliferation of nuclear weapons been effective?
This question refers to the ongoing increase in nuclear explosive capacity among the great powers. The answer is not favourable. The nuclear explosive capacity of the superpowers has reached such enormous proportions that it is difficult to understand why the efforts at vertical non-proliferation have not been effective. By way of illustration of the situation, since the first nuclear explosion the stockpile of nuclear weapons has reached an explosive power equivalent to 1 300 000 bombs like the one dropped on Hiroshima. While there are only five NWS, they have not only increased their explosive power and capacity, they have also increased and perfected the vehicles carrying those weapons, the accuracy of their missiles and their aircraft, since the start of the atomic era.

Has horizontal non-proliferation of nuclear weapons been effective?
The international community is aware that if the number of NWS increases, the probability of a nuclear war increases proportionately. For this reason efforts have been made to stop the proliferation and apply measures to prevent it, starting with the IAEA in Vienna. The aim there was to control the supply of fissionable and nuclear materials so that they were

used exclusively for peaceful purposes and not diverted for warlike ends. Subsequently the NPT was undertaken.

A final point here is that horizontal non-proliferation efforts have been more successful by far than those related to vertical non-proliferation.

Has geographic non-proliferation of nuclear weapons been effective?
Nuclear weapons have been disseminated outside the territories of the NWS to different geographic areas. Moreover, the use of nuclear submarines poses the risk that a nuclear war could be started anywhere.

It must be remembered that nuclear weapons are daily located in the water and air space of the NNWS. The Mediterranean, Caribbean, Indian Ocean and other seas are frequently used for military manoeuvres. This situation is in fact geographic *vertical* proliferation, that is, a nuclear threat.

What has been the reaction to this phenomenon? Have the civil and military facilities of the NWS been subjected to safeguards so as to bring the freedom to test, develop, perfect and deploy new weapons under efficient international control?

Relation Among Non-Proliferation Efforts

Efforts at non-proliferation in any of its dimensions must be performed in a parallel manner; however, they must also be convergent, since they are aimed at a more stable world. They must not be mutually subordinated. Advantage must be taken of all favourable circumstances for reaching a significant reduction in all types of nuclear arsenals.

The fact that the leading powers in both great alliances have a special responsibility for nuclear disarmament is not under discussion. However, that responsibility is not exclusive, since life on earth depends on it. Non-proliferation is every nation's responsibility, whether nuclear or not, as is shown by the creation of nuclear weapon-free zones.

EFFORTS AT NON-PROLIFERATION OTHER THAN UNDER THE NPT

Efforts at horizontal non-proliferation have not been limited to the NPT. They have included the establishment of nuclear weapon-free zones, the safeguards agreements signed by IAEA member states, and other unilateral and bilateral agreements such as those between Argentina and Brazil. Other measures besides the NPT that have been applied include: the creation of nuclear weapon-free zones; security guarantees have been made to the NNWS; the prohibition of nuclear weapons tests; the freezing of the production of nuclear weapons and of fissionable material; a general agreement on non-use or non-first use of nuclear weapons; and concrete disarmament measures.

Nuclear Weapon-Free Zones

The definition of a nuclear weapon-free zone is found in paragraph 33 of the final document:

> the creation of nuclear weapon-free zones on the basis of agreements or arrangements reached freely by the countries of the respective region and the full compliance with those agreements or arrangements, thus ensuring that the areas be kept really free from nuclear weapons, as well as the respect to those zones by NWS, constitute an important disarmament measure.

Tlatelolco Treaty

This treaty, which was signed in 1967 before the NPT, established a nuclear weapon-free zone in Latin America. The twenty-five countries involved in the treaty promised not to test, produce or acquire nuclear weapons, nor to allow any other power to do so or to locate or deploy such weapons in their territories. According to Protocol I of the treaty, those non-Latin American countries controlling some territories located within the boundaries of the zone of the treaty – France, the United States, the Netherlands and the United Kingdom – would agree to consider those territories as part of the nuclear weapon-free zone. Except for France, which will make a decision in due time, the other nations have already signed it.

According to Protocol II, all the NWS promised to respect the nuclear weapon-free zone status of Latin America and not to use or threaten to use nuclear weapons against the parties of the treaty.

For its part, Argentina has always shown permanent support for the spirit and goals of the treaty. It provided clear evidence of this attitude when it participated in the elaboration and signing of the treaty. However, Argentina has serious objections to Tlatelolco which are hindering its ratification of the treaty, such as:

1. The protection of industrial secrecy, which would be compromised by the control functions of OPANAL, especially in the case of the special inspections that are carried out at the request of any

member country (see Articles 12 through 18). This issue is of especial concern to Argentina because of its recent technological progress, which has been attained without foreign assistance. The Tlatelolco Treaty poses a serious risk since it does not entail a confidentiality system. Breaches of confidentiality also represent an important proliferation threat.

2. The lack of agreement on a safeguards model specific to the Tlatelolco Treaty and the IAEA's interest in having the same type of safeguards apply as were established for the NPT.

3. The lack of credibility of the NWS with respect to their meeting their obligations under additional protocols, since the treaty has no verification system to assure they meet those obligations.

4. The presence of British nuclear weapons in the South Atlantic, which the United Kingdom has not categorically and fully denied. It should be noted that in October 1982, when referring to the Falkland Islands and the presence of nuclear weapons on the British fleet in the South Atlantic, the spokesman of the House of Commons stated that 'It would not be in interests of national security to depart from longstanding practice, observed by successive governments, neither to confirm nor deny the presence or absence of nuclear weapons in any particular place at any particular time.' If, for national security reasons, a country may hide the presence of nuclear weapons in areas supposedly free from those weapons, what is the worth of promises made to respect the nuclear weapons-free zones? The experience in the South Atlantic relative to the Tlatelolco Treaty is useful and should be analysed per Article 1 of the treaty, which explicitly prohibits the 'receiving, storage, installation, settlement or any other way of possession of any nuclear weapons'.

5. The interpretive statements made by the signatories and the wording of parts of the additional protocols, about which Argentina has real reservations as they modify the wording of the treaty itself while introducing discriminatory elements. An example is the unilateral statements made by powers with reference to the non-use of nuclear weapons against the NNWS. The statements are constantly being submitted with conditions so that for all practical purposes they preclude any guarantee.

6. The incompatibility between the Argentine position concerning the transport of nuclear weapons in a military denuclearized area and the reservations of the NWS allowing their transport. What is the use of a nuclear weapons-free zone if the countries that should be the first to promise firmly to respect the zone are the ones that

in fact are free of any possibility of verification and maintain the broadest freedom to introduce weapons into the area and to hide that behaviour for national security reasons? What degree of tranquility and security can the other countries in the zone, which are subject to verification, have in these circumstances? If the rights of the NWS are not coupled with obligations such as those of the NNWS, it may be wondered what the use of a denuclearised area is to the NNWS members.

The Rarotonga Treaty, 1985

This treaty establishes a nuclear weapons-free zone in the South Pacific, now the second area free from nuclear weapons. The nine members of the South Pacific Forum that ratified the treaty on 31 July 1987 promised not to manufacture, acquire or control by any means explosive nuclear devices within or outside the zone and to prevent the stationing and testing of any nuclear explosive device. In addition, the treaty prohibits the pouring of radioactive wastes into the sea or any place within the zone. The treaty does not say anything concerning the freedom of navigation according to international right.

According to Protocol I, the United States, France and the United Kingdom would promise not to manufacture, station or test explosive nuclear devices in the territories under their responsibility located within the zone. They have not yet signed that protocol.

According to Protocol II, the five NWS would promise not to use or threaten to use any nuclear explosive device against the parties of the treaty or against any territory within the zone under the responsibility of the NWS parties involved in Protocol I. By late 1987 China and the Soviet Union had signed the protocol.

Regarding Protocol III, it bans the testing of explosive nuclear devices within the zone. It, too, has not been signed by the five NWS. By late 1987 China and the Soviet Union had signed this protocol.

This treaty came into force 11 December 1986.

Safeguards by EURATOM

This agency, created after the 1957 Treaties of Rome, has its own safeguards system to ensure the peaceful use of nuclear energy. Following difficult negotiations between EURATOM and the IAEA, an agreement was reached in Brussels on 5 April 1973, which came into force on 22 February 1977. The two safeguards systems were made compatible by creating a system that is applicable to the NNWS that are members of

EURATOM. The territories of member countries are considered as a single country. That is, transfers of nuclear material or equipment among them are not considered international transfers. Japan has, same as EURATOM, an agreement with the IAEA that has some different characteristics. Thus the NPT contains yet another instance of discrimination.

Unilateral Measures

Several countries considered to be 'threshold' states have taken unilateral measures to observe non-proliferation principles. In this regard, I cite the agreement for 'Argentine-Brazilian Cooperation concerning the peaceful use of nuclear energy, within the framework of the integration process'.

Cooperation between Argentina and Brazil on the peaceful application of nuclear energy has been both intense and long-lived. They signed their original agreement on 17 May 1980. Previously they had co-operated in terms of industrial, professional, scientific and technical exchanges in the nuclear field. This cooperation established a working relationship that became stronger as time went by and in turn allowed each country to gain considerable knowledge regarding the activities of the other.

An important event in their bilateral relations was the Declaration on Nuclear Policies, signed by their presidents on 30 November 1985 at Foz de Iguazú. The declaration had three main objectives:

1. Securing and strengthening bilateral cooperation.
2. Overcoming the restrictions imposed by countries advanced in the nuclear field that were hindering technological and industrial development by Argentina and Brazil.
3. Reducing the risk of an arms race in the region.

Also, within the framework of the declaration it was decided to:

> create a joint working group, under the responsibility of both Chan-cellorships, integrated by representatives of the respective Nuclear Commissions and enterprises, aimed at promoting relations between both countries in that area, the promotion of nuclear technological development, and the creation of mechanisms to ensure the superior interest of peace, the security and the development of the region . . .

As a result of this provision a Permanent Committee was appointed. Convening every three months, the committee has already held seven working meetings. Within the committee are three sub-groups analysing various aspects of bilateral nuclear cooperation. Their conclusions are subject to evaluation during the plenary sessions.

Under Protocol No. 11 (Act for Argentine-Brazilian Integration, Buenos Aires, 29 July 1986) and its Annexes I and II (Brasilia, 10 December 1986), Argentina and Brazil promised to provide immediate information and reciprocal assistance in the event of nuclear accidents or radiological emergencies. Still other areas of cooperation were defined in the field of nuclear safety and radiation protection. It must be noted that Protocol No. 11 was invoked after the accident at Goiania, Brazil. Brazil supplied information to Argentina, Argentine experts offered 'in situ' advice, clinical analyses performed in Brazil were rechecked in Argentina, and Argentina provided medical assistance.

The two countries also made their external nuclear policies compatible and elaborated common strategies for action in several international fora and in dealings with third parties. Thus, Brazil and Argentina, which started with traditionally concordant positions, are now able to respond better to the pressure that both countries bear because of their autonomous programmes of nuclear development.

Argentina and Brazil have created a 'Sub-group of Legal and Technical Requirements for Cooperation' that has furthered the objectives included in Article IV of the NPT on safeguards agreements with the IAEA and the juridical aspects involved in nuclear exports and imports to and from third parties. In addition, they are analysing the juridical implications of their own exchanges of materials and equipment.

It must also be noted that within the framework of this process the two countries have begun a process that has great international application and that has created a climate of growing mutual confidence and transparency concerning the respective nuclear programmes. It involves reciprocal visits to nuclear facilities. Joint presidential visits have already taken place at the Argentine uranium enrichment plant at Pilcaniyeu (July 1987), the Brazilian uranium enrichment plant at Iperó (São Paulo, April 1988) and the Argentine reprocessing plant at Ezeiza (November 1988).

Partial Test Ban Treaty

This treaty, which was signed in Moscow in 1963, bans nuclear weapons tests in the atmosphere, outer space and under water. Among the NWS that have not signed the treaty are France and China. The former says it will do so when a general agreement is concluded banning nuclear weapons and eliminating existing ones. The latter has adopted a similar attitude.

This treaty is evidence of the concern generated by the diffusion of nuclear weapons and the radiation generated during each test. The objectives are' . . . the speediest possible achievement of and agreement on general and complete disarmament under strict international control; and an end to radioactive contamination of the environment'. Article I of the treaty prohibits 'any nuclear weapon test explosion, or any other nuclear explosion' in any of the aforementioned environments.

The Antarctic Treaty

This treaty was signed in Washington, DC in 1959. Article I states that 'Antarctica shall be used for peaceful purposes only'; and that to that end, inter alia, 'the testing of any type of weapons' will be forbidden. Article V adds that 'Any nuclear explosions in Antarctica and the disposal there of radioactive waste material shall be prohibited.' The verification system specified in the treaty is being carried out. The breadth of the system is unprecedented, as may be seen from Article VII, paragraph 3. This treaty is preserving an important area of the planet free from nuclear weapons and their impact.

The Sea-Bed Treaty of 1972

This treaty, which seeks as a secondary goal general and complete disarmament under international control, constitutes an important step toward excluding the sea bed, ocean floor and subsoil thereof from the arms race. This point is made explicitly in the third and fourth paragraphs of the treaty's preamble and in Article V, where the parties agree to continue negotiations toward that objective. Logically, as a treaty dealing with non-armament of the sea bed and ocean floor, it involves a measure against the proliferation of nuclear weapons. The difficulty with this treaty is that most countries do not have the means or the capacity to verify violations. Geographic proliferation through the use of nuclear-equipped ships on the surface and under the water complicates the situation.

Guidelines by the London Club

At the initiative of the United States, exporters of nuclear technology held a meeting in London to establish more severe conditions for foreign trade and to prevent export competition from weakening the safeguards. After the meeting the London Club, as the group was called, issued the 'London Guidelines', approved September 1977. The London Club originally had seven countries as members. Presently there are fifteen

(Belgium, Canada, Czechoslovakia, United States, France, Italy, Japan, the Netherlands, Poland, the United Kingdom, the German Democratic Republic, the Federal Republic of Germany, Sweden, Switzerland and the Soviet Union). Other countries that respect the guidelines are: Australia, Denmark, Finland, Greece, Ireland and Bulgaria.

The Treaty on Outer Space

This Treaty deals with another environment that the signatories are trying to keep free of weapons proliferation. The treaty was signed in Washington, London and Moscow in 1967. Article IV states that 'States Party to the Treaty undertake not to place in orbit around the earth any objects carrying nuclear weapons or any other kinds of weapons of mass destruction, install such weapons on celestial bodies, or station such weapons in outer space in any other manner.' The treaty makes no mention of a nuclear missile traversing space.

The 'negative guarantees'

'Negative guarantees' are efficient international arrangements that provide guarantees to the NNWS against the use or threat of use of nuclear weapons. The question is whether a NWS that believes its national security is being vitally threatened can be trusted not to use all the means available to defend itself. To guard against that possibility the most efficient guarantee consists of the 'prohibition and later destruction of all nuclear weapons'.

'Unilateral statements', each one based on a country's own strategic perceptions, are not alike. They contain a series of conditions, requirements and escape clauses that so diminish their value as a guarantee that the security for a NNWS is almost non-existent.

In terms of unilateral measures aimed at ending both horizontal and vertical proliferation and taking advantage of the improved international climate of *rapprochement*, the NWS should issue new policies that will give the NNWS confidence, lead to nuclear disarmament and guarantee peace through the elimination of nuclear arsenals. The nuclear powers must face the non-proliferation challenge and reverse the arms race. The future will depend on their imagination and flexibility in addressing those goals.

The People's Republic of China has categorically promised that under no circumstances will it ever be the first nation to use nuclear weapons, nor will it ever and unconditionally use them against the NNWS. The USSR has made the same promise.

OTHER GENERAL MATTERS CONCERNING PROLIFERATION

The industrialized and developed countries have almost all the basic requirements and scientific and technical competence needed to design and build the simplest nuclear weapons. However, to build those weapons, they must deal with other non-technical and economic barriers. They need to obtain explosive nuclear materials, such as highly enriched uranium or plutonium, which in turn requires highly complex facilities or the capital for their installation. To apply safeguards and avoid the diversion of materials for other purposes a series of requirements must be fulfilled that may or may not be covered by the IAEA or other types of safeguards agreements. It is important to note that the programmes involving uranium enrichment or fuel reprocessing plants may also be used as sources of nuclear materials for peaceful purposes.

Another important issue is the transfer of technology, equipment and scientific knowledge aimed at encouraging the development of the peaceful use of nuclear energy, as specified by the NPT. Many countries use the potential for dual use of such knowledge as an excuse not to permit transfers.

The safeguards system presently in use is discriminatory, as the countries are subject to different rights and obligations based on the mere fact of their having been 'nuclear' weapon states on a given arbitrary date. Moreover, no special consideration has been given to industrial secrecy and free marketing in an equitable manner. The latter refers to the ability of new suppliers of nuclear technology to enter the market, a development opposed by the exclusive London Club and the guidelines of these long-time suppliers.

The continuous multiplication of sophisticated nuclear missile systems and the presence of ships capable of deploying nuclear weapons pose a risk of a nuclear war. Avoiding the proliferation of nuclear weapons is an integral – and essential – part of the total and complex process of disarmament and must be viewed jointly with all the other issues related to nuclear disarmament, and not as an isolated issue.

Let me then quote paragraph 56 of the final document of the First General Assembly on Disarmament: 'the most efficient guaranty against the danger of a nuclear war and against the utilization of nuclear weapons is nuclear disarmament and a complete elimination of nuclear weapons'. It is also important to remember the statement made by the heads of state of the non-aligned countries, who met in Delhi in 1983:

the renewed intensification of the nuclear arms race, both quantitatively and qualitatively, and confidence in the doctrines of nuclear dissuasion, have increased the risk of a nuclear war and have brought along greater insecurity and greater instability in international relations. Nuclear weapons are something beyond warlike weapons. They are mass annihilating instruments.

That quote explains why all the theories and concepts about the possession of nuclear weapons and their utilization under any circumstances were rejected.

However, can we deal with nuclear proliferation, nuclear disarmament or the cessation of the nuclear arms race, with the proliferation of chemical or radiological weapons and with the prevention of weapons proliferation in outer space separately? Of course not! We understand that the prevention of the proliferation of nuclear weapons, either vertical or horizontal, cannot be considered as separate from the prevention of war as a whole and from issues related to the maintenance of peace.

By way of a conclusion for the comments on non-proliferation, from a point of view that reflects serious doubts concerning the philosophical support for the NPT and its applicability, I note that any type of proliferation – vertical, horizontal or geographic – is condemnable in itself.

Let me finish with the words of Ambassador Major Britt Theorin:

We need these actions now! We need them now, as the time is running away from us! We need them now, and they are possible now. We need them now, because a breakthrough in these areas would be a strategic push to go further on the road to total disarmament, visionary justice, sustainable development and lasting peace. We owe these actions to the victims of the nuclear bombs over Hiroshima and Nagasaki. We owe them to all the undernourished and starving, to the poor and uneducated, to the sick and homeless of the world. We need to stop the arms race on all fronts. We owe these actions to our children and their children and their unborn grandchildren.

Response – *Edmundo Fujita*

I wish to identify myself with the views expressed by Martin Gomez Bustillo. I do so not just personally. The convergence of views between Brazilian and Argentine diplomacy in the field of disarmament predates our recent efforts at cooperation in the nuclear field. This convergence of views goes back to the 1960s, to the discussions that ultimately led to the negotiation of the Nuclear Non-Proliferation Treaty (NPT). From the very start our two countries shared common views. Together we identified the flaws and defects of the treaty and the consequences that would arise therefrom. Nowadays the voting patterns of Brazil and Argentina in the United Nations First Committee or the Conference on Disarmament are virtually identical. Therefore I do not have much to add to Mr Gomez Bustillo's presentation, which was very complete. What I would like to do is to focus on a few points I think are relevant to a discussion of the non-proliferation question.

In his paper Mr Gomez Bustillo made very clear the different dimensions of non-proliferation – the vertical, the horizontal and the geographic. I would like to stress a curious semantic reductionism that tends to distort discussions of non-proliferation and that diverts attention from the real dimensions of the problem. In fact, the majority of the literature in this field tends to focus on horizontal proliferation and, in general, to touch only lightly on vertical proliferation.

I would like to suggest that there is a phenomenon I call the 'ideology of non-proliferation'. By ideology I mean a certain simplification of vision, a pseudo-consciousness of the problem that may lead to a misrepresentation of reality. This ideology may reflect the interest of certain hegemonic groups in freezing the status quo not only in the military and political fields, but also in the scientific, technological and economic fields. This ideology may be spread by many means, overt or subliminal, and may end up convincing wider groups, which then proceed to incorporate the views of such hegemonic groups.

I would like to point out certain characteristics of this ideology of non-proliferation. As I said, it is a form of semantic reductionism. It refers more to horizontal proliferation than to vertical and geographic proliferation. It has a certain Manichean vision of the world. It says that you have the good side – those who signed the NPT, perhaps – and the bad side – those that did not sign. This kind of view makes a messianic claim to moral superiority, and to the notion that the good side knows best how to make the world safer and can teach others what they should do.

Another characteristic of this ideology is that it is all-encompassing. What began as a problem of preventing the horizontal proliferation of nuclear weapons now extends to nuclear technology, missile technology, chemical weapons and who knows what else in the future.

This ideology is implemented by a 'non-proliferation regime' that is essentially based on the Nuclear Non-Proliferation Treaty (NPT), the IAEA's full-scope safeguards, the London Supplier Club's guidelines and, perhaps, some new addition such as the Missile Technology Control Regime.

How is it possible to grasp the true reality? With respect to vertical proliferation it is important to note that forty years ago there were no nuclear weapons. Today we have over 50,000 nuclear warheads deployed worldwide. As for horizontal proliferation, there are no ascertainable signs it has occurred.

In 1974 India exploded a nuclear device. Since that time India has not developed into a nuclear weapons state, at least not ostensibly. Although there are strong suspicions that certain countries may be developing nuclear weapons, other countries that clearly do not have a nuclear weapons programme are nonetheless caught in this wide web of suspicion. There is a tendency to group together all those nations that have not signed the NPT, for example. I would emphasize that Brazil and Argentina can in no way be compared with South Africa or Israel or India or Pakistan. The only rivalry between Brazil and Argentina that I am aware of is in soccer.

How can this non-proliferation ideology be countered? First, we should restore the original, integral meaning of non-proliferation. I refer to the linkage that was mentioned in the title of this panel – the structural linkage among vertical, horizontal and geographical forms of proliferation. Instead of trying to chase after a few prospective proliferators, we should endeavour to stop the proliferation that now exists.

We must try to avoid repeating past mistakes that have allowed loopholes in the treaties that seek to eliminate nuclear weapons. I am speaking of loopholes that permit new, even more sophisticated armaments to compensate for reductions in existing weapons systems. Further, I refer to the loopholes that permit the deployment of weapons in new geographic areas and even in outer space to compensate for weapons that are withdrawn.

We should ask the nuclear weapons states to take concrete measures to reduce their arsenals. Both sides – East and West – should abandon their doctrine of deterrence. If the deterrence theory is correct, if it has indeed kept the peace for the past forty years as the nuclear weapons states claim, then they are giving other countries an incentive to adopt the same theory. This position is absolutely insane.

Pending the complete destruction of the arsenals of the nuclear weapons states, we could request they give clear, unambiguous and unconditional negative security assurances to the non-nuclear weapons states. Furthermore, we could ask that they strictly refrain from introducing nuclear weapons into nuclear weapons-free zones and zones of peace. The nuclear weapons states must abandon their practice of refusing to disclose whether their ships and submarines are carrying nuclear weapons. They must submit to strict measures by non-nuclear weapons states to verify that no nuclear weapons are being carried into these regions. Finally, they must withdraw all reservations and interpretive declarations that they have made in treaties such as Tlatelolco.

In addition, they should not impose obstacles to the development of nuclear energy for peaceful purposes. Instead, they should co-operate fully with those non-nuclear weapons states that seek to develop a civilian nuclear industry. They should reduce the inordinate amount of financial and technical resources they now spend on the development of new weapons of mass destruction and should instead redirect those resources to assist in the development of the Third World.

In conclusion, I would like to say that these are some common sense observations that non-nuclear weapons states have addressed to the nuclear weapons states over the years. So far we have seen no tangible results of the weapons states' commitment drastically to reduce and ultimately to eliminate nuclear weapons.

The final document of the UN General Assembly's First Special Session on Disarmament identifies nuclear disarmament as the highest priority of the international community. This authoritative document was adopted unanimously by the international community, including the five nuclear weapons states. It calls on those who do not have nuclear weapons to refrain from acquiring them and on those who do have them, to eliminate them.

The non-nuclear weapon states have done their part. We hope the nuclear weapon states will do their part, too.

Discussion – *Rear Admiral Thomas Davies, USN (Ret.)*

To understand this area one needs to review briefly the history and nature of arms control and arms control treaties. In an earlier era many treaties and agreements were labelled 'arms control'. For example, there was the Kellogg-Briand Treaty to reject the use of force in foreign policy as a means of achieving national objectives. Another was the 1926 Geneva Convention, in which the signatories agreed not to use chemical or biological weapons. These treaties became meaningless over the years because they were 'hortatory'. They were empty words, without meaning except to ensure short-term political gains for local politicians. Yet they were given great credence and were published around the world.

In more recent times arms control has taken on a different complexion. We have begun to realize that, for a treaty to be effective, it must eliminate or prevent the acquisition of a specific class of weaponry. To say in advance that we are not going to use a weapon is hortatory and meaningless. Not to have a particular weapon available to use is a different story. This is the way modern arms control treaties should be constructed, we have agreed.

We find, however, that these treaties are not as easy to negotiate as the old hortatory ones. The Congress of the United States ratified the Kellogg-Briand Treaty in about an hour. There was no debate because the treaty did not mean anything. Limiting the number of warheads or missiles that each side can build is, however, a more difficult decision. First, members of Congress must understand the issues. Second, they must agree to renounce the use of a particular weapon and therefore decide not to have it available. Either we will destroy the weapon, as in the proposals on chemical weapons, or we will destroy the delivery system, as in the case of the INF treaty. This step is quite different from ratifying a hortatory treaty. It must be taken with the greatest seriousness and only after lengthy debate. We have moved very slowly in this direction.

When we talk about non-proliferation, I think we are now speaking, in this new context, of our intention to make certain that nations do not have the means to build a nuclear weapon. This is the spirit behind the Nuclear Non-Proliferation Treaty. If we are looking at non-proliferation in a modern context – as a way to prevent the construction of nuclear weapons or, in the case of vertical proliferation, to do away with the weapons that we have – these are steps that must be taken very cautiously in a democracy.

There is also the problem of verification. I have heard the words 'mutual trust' used a great deal here. The Soviets used to dwell on mutual trust in our negotiations on the Comprehensive Test Ban Treaty. They would say verification was not a problem because there was mutual trust. My colleague Paul Warnke answered them once by saying, 'I trust the Soviet Union completely to look out for its own interests, and I hope they feel the same way about the United States.'

Concerns about verification lead to a discussion of safeguards and the other devices we have discussed here in the last two days, all of which go to the issue of serious, practical execution of an arms control arrangement. Any nation – and I am sure it would be true of Brazil and Argentina – placed in the same position as the United States would demand the same results, that there be some sort of effective verification.

Verification measures should be kept as simple as possible. No one likes bookkeeping, although I have heard it said here that a national bookkeeping system for nuclear material is essential, just to keep track of what you are doing. In the United States there have been incidents in which nuclear material has been missing. The media has used the US Freedom of Information Act to find out about the missing nuclear material, to the great embarrassment of the government. These incidents have been made public. In the end, these revelations have turned out to be good because they have helped to stimulate corrective action.

Publication of data and information on the control of your nuclear materials, difficult though it might be, would be useful – we call it the 'sunshine' approach. In the end I think you will find that this approach produces good results.

Let me stress that none of us is in a position, either juridically and certainly not from the standpoint of knowledge, intelligence or righteousness, to tell any other country what to do. However, I do believe it is interesting to think in terms of the arms control constructs I have just mentioned. We have gone from the hortatory – and I consider negative security guarantees to be hortatory – to pragmatic measures where we actually relinquish some of our sovereignty. When we do so, it takes a long time, it is a serious matter, and verification must play an important part.

I want to make a few comments about submarines.

Although I am a naval aviator, I have worked with submarines a great deal. In fact, during the Cuban missile crisis the surveillance I directed in the Atlantic consisted of aircraft and submarines working as a unit. In addition, my father-in-law was a submariner, so that I always had to be careful to talk about submarines with great diplomacy.

The advent of anything nuclear always produces an emotional reaction that sends common sense flying out the window. It has happened in the case of nuclear submarines. There are many people in the US Navy, myself included, who have thought for a long time that we should be building conventional submarines as well as nuclear submarines. There is a case to be made for both. For example, a conventional submarine is much quieter than a nuclear submarine. This difference is obvious if you stop to think about the propulsion systems. There are situations where nuclear submarines are superb and well worth their enormous cost. There are also cases where you could do as well or better with the conventional boats.

Technology has a rather shocking impact, an emotional impact. 'Glamour' is a word used by some of my colleagues here. It carries people away. I think of the MIRV, the multiple independently targeted warhead, a re-entry vehicle 'invented' during the SALT I negotiations as a counter to the anti-ballistic missile. At one time during the negotiations, after the MIRV had been invented on paper at least, it was proposed that we ban the MIRVs in the SALT I agreement. Our own technical people were so emotionally stirred by this wonderful new idea that we did not move to ban them, and the 50,000 warheads on both sides you have referred to are the result. Dealing with nuclear technology requires a particular kind of realism that is hard to come by.

7 Global Significance of a Bilateral Arrangement between Argentina and Brazil

Presentation – *Minister Roberto García Moritán*

REFERENCE FRAMEWORK

One of the priority needs in Latin America is to generate sufficient capacity through different sources of energy to improve the population's standard of living, expand the region's industry and technology and generate sustained economic growth. Within this framework nuclear energy plays a key role in the development programmes of at least two of the countries in the region: Argentina and Brazil.

Among other benefits nuclear energy, in addition to generating electric power, allows for the production of radio-isotopes that are used in agriculture, medicine and industry, for the improvement and expansion of food production and for the creation of a local technological base liable to stimulate scientific development. These benefits led the developed countries to move quickly and with a clear conscience to develop all the peaceful applications of nuclear technology. Developing countries at a certain technological level also moved toward nuclear power to satisfy their needs, diminish their economic imbalances and overcome their underdevelopment.

This trend toward the use of nuclear technology has political and commercial ramifications that have precipitated efforts by some countries – under the guise of non-proliferation of nuclear weapons – to limit the access of developing countries to certain areas of the technology. In reality they have been trying to maintain their oligopoly over the supply of technology for peaceful applications of nuclear energy. Such restrictive stances (which have taken the form of common measures, principles and guidelines) have produced serious concern and inconvenience among the developing countries, obliging them to invest resources in attaining self-sufficiency that otherwise would be unnecessary, to secure a reliable, safe and permanent supply of the nuclear fuel for their nuclear power plants.

Self-sufficiency was the reason Argentina decided to pursue a nuclear programme aimed at the development of all stages of the fuel cycle, from mining uranium to disposing of radioactive wastes. Its programme includes the production of uranium, the fabrication of fuel and the reprocessing of spent fuel. Another priority is to supply zircalloy tubes for the manufacture of the fuel rods and heavy water used as a reactor moderator. Argentina made a decision to develop uranium enrichment technology when the country supplying the fuel for its research reactors interrupted its shipments. Without that supply Argentina would have been unable to comply with promises it had made to co-operate with countries in the region.

The Pilcaniyeu plant, with a capacity of 20 000 units of separation work is to be expanded to 100 000, which will allow the use of slightly enriched uranium in the heavy water reactors and thereby reduce the cost of nuclear generation and extend the life of Argentina's uranium reserves. This plan and the decision to recycle transuranic elements will lead to the optimum use of reserves.

It is a pity that regions such as Latin America, which have so many economic and social priorities, are obliged to diffuse their efforts and resources to achieve autonomy of power generation. However, without that autonomy they would be dependent, a status that would seriously compromise their development plans.

These matters call for international consensus on access to nuclear technology based on the principles of non-discrimination, free access to the technology and the right by all countries to start nuclear programmes for peaceful purposes in accordance with their interests, needs and priorities.

Countries such as Argentina and Brazil, in response to the present trend among the traditional suppliers of nuclear technology, are placing more and more importance on South-South assistance and cooperation. Through coordinated action within the United Nations and other international organizations, especially the International Atomic Energy Agency (IAEA), they have pointed out the political nature of these implicit problems and the corresponding need to apply a political dimension to the search for consensus.

NUCLEAR COOPERATION BETWEEN ARGENTINA AND BRAZIL AS A BASIS FOR CONFIDENCE-BUILDING

During the 1970s and early 1980s both Argentina and Brazil made important progress in their respective nuclear development programmes, both of which are aimed at meeting local needs. Initially some traditional suppliers played a leading role in that development. However, during that period those powers came to take a much harder line, and made the transfer of certain technologies harder and more restrictive. As a result Argentina and Brazil decided to gain command of those technologies on their own. In parallel with this situation the local nuclear industry in both countries was growing in keeping with the countries' ongoing development and future projects.

It must be recognized that during that long period neither Argentina nor Brazil considered co-operating with one another in the use or applications of nuclear technology. To the contrary they selected technologies based on internal considerations and the dynamics of the technologies rather than on what the other country was doing. The delay in starting to integrate nuclear development and the lack of experience with doing so allowed supplier countries to profit from the situation.

In 1980, however, the two countries took an important step by signing a cooperation agreement for the peaceful use of nuclear energy. Within that context they initiated technological cooperation. A few years later they took a second important step. The two countries provided remarkable political support to bilateral relations in the nuclear field, trying to extend and develop it in every aspect.

In sum, the present structure of bilateral relations has been built step by step. Within the framework of the cooperation agreement we have put in place a joint programme that involves periodic meetings and technical visits, analysis of joint projects, evaluation of the feasibility of complementary activities by our nuclear industries and the like. This process has created a network of reciprocal interests encompassing both technical and political considerations.

Both countries see the process as the start of a long-lasting relationship. A system has been structured that is unique among developing countries in the same region at a similar level of nuclear development: there is no nuclear facility in Argentina and Brazil that has not been visited by technicians from the other country.

It is useful to review the objectives of the Foz de Iguazú declaration on cooperation:

. . . nuclear science and technology are transcendental in the life of a modern country, since they significantly promote its social and economic development;

. . . both countries have made strong efforts, during many years, per-

forming research and studies concerning the peaceful applications of
nuclear energy; this has required important investments, in order to
reach a level of knowledge allowing them to provide the countries'
inhabitants with the benefits arising from the peaceful use of nuclear
energy;

. . . cooperation between Argentina and Brazil will constitute a factor
multiplying the benefits that they may obtain reciprocally from the
peaceful use of nuclear energy, thus allowing both countries to attain
better conditions concerning the difficulties in the international supply
of nuclear equipment and materials;

. . . such cooperation will be open to all Latin American countries
interested in participating.

The declaration also reiterates the promise to develop nuclear energy
exclusively for peaceful purposes, points out the purpose of close cooperation in all the fields related to the peaceful use of nuclear energy, and states
the hope that cooperation will be extended to other countries in the region
that share the same objectives.

The Foz de Iguazú declaration established a working mechanism
involving the chancellorships, national atomic energy commissions and
nuclear enterprises of both countries. The aim is 'the promotion of nuclear
technological development and the creation of mechanisms ensuring peace,
security and local development, without disregarding technical matters in
nuclear cooperation that will continue to be ruled by the regulations in
force'.

Three years later the presidents of Argentina and Brazil signed the Iperó
Declaration. It deals, among other things, with the following: 'The fact that
bilateral cooperation in the nuclear field has brought along new ways of
collaboration has, in turn, through contact at a political and technical level
and a significant exchange of information, contributed to the consolidation
of mutual confidence.' Emphasis is also placed on 'significant events, such
as President Sarney's visit to the uranium enrichment plant in Pilcaniyeu'
and the official Argentine visit to the Aramar Experimental Centre at Iperó,
plus supplemental technical visits in both cases.

MATTERS DESERVING ATTENTION

Analysis of the ways in which the various issues posed by nuclear technology have been dealt with in the various international fora, governmental

and other, reveals a restrictive approach both substantively and instrumentally. As far as the *substantive matters* are concerned, the emphasis has been narrow, focused exclusively on avoiding an increase in the number of nuclear weapons states. Examples of this tendency are:

1. The Nuclear Non-Proliferation Treaty, which, by legitimizing the right of a small group of countries to base their security on the use of nuclear weapons, consolidates an oligopoly among the suppliers of that technology.

2. The creation of the so-called 'London Club', a commercial arrangement under which the members, under the guise of contributing to the prevention of horizontal proliferation of nuclear weapons, issued a list of equipment and facilities that could not be transferred. This list went beyond the restrictions of the NPT, including items that do not have potential military application. They do, however, seriously affect the possibility of developing important peaceful applications of nuclear technology and thus of providing people with important benefits.

At the same time the club does not give equal weight to some important matters, such as the start of widespread cooperation furthering the peaceful use of nuclear energy overall and the need to end the quantitative and qualitative increase in the nuclear arsenals of the beneficiaries of the NPT and their military and commercial allies.

Concerning the *instrumental matters*, first, a universal link was made between the concept of control and the use of a safeguards system, specifically that created by the IAEA in INFCIRC/66 Rev. 2, approved in 1968. Later, when the IAEA's safeguards system was adapted for application to the NPT, the idea of effective control by the system known as 'fullscope safeguards', as shown in INFCIRC/153, was erroneously associated with the abolition of the original IAEA mechanism known as 'case-by-case safeguards' mentioned in the above paragraph. Thus, the concept of control gradually came to be limited to 'safeguards,' which concept in turn, although embodying broad connotations in its current meaning, was limited to a certain type of control and, within that type of control, to a given mechanism.

It is interesting to remember the response of a group of European countries to the above system. They developed a different control system that accommodated their autonomous decisions, technological evolution and individual needs, such as the preservation of industrial secrecy. Inclusion of the provision in Article III of the NPT that states could negotiate their safeguards with the IAEA in collective form was a response to their concerns. It should also be remembered that negotiation of the safeguards agreement between the IAEA and EURATOM took a long time, only coming into force in 1977. Even then, Japan was later able to obtain special treatment by the IAEA in terms of the features of its safeguards agreement. In Latin America the treaty banning nuclear weapons in the region and in the Caribbean was a pioneering effort that has suffered from the effects of technological evolution and the restrictive interpretations of some traditional suppliers. An instrument that originally was not discriminatory has been distorted so that it now uses the concept of control to accomplish its objectives in an inappropriate way.

The failure of the scheme promoted by the NPT has been exacerbated by the behaviour of the IAEA regarding those countries that, according to Article XIII, promised to negotiate a safeguards system with the IAEA in accordance with the wording of the treaty. In fact, using a barely justifiable practice the IAEA adopted a negative attitude toward the negotiation of safeguards agreements under the Tlatelolco Treaty and attempted instead to apply the safeguards system designed for the NPT in every case. This tactic adulterated the spirit of that regional instrument and erroneously integrated both treaties.

This situation deserves analysis, since, through its instrumental approach, the IAEA, which is supposed to grant assistance, in fact, is imposing requirements and formulating interpretations that go beyond the intent of the Treaty, modifying the obligations the contracting parties originally assumed. The important differences in the IAEA's attitudes toward the European countries and Japan, on the one hand, and toward Latin America, on the other, are remarkable.

For the above reasons the NPT/IAEA system cannot be analyzed independently from the behaviour of the London Club, from its restrictive policies concerning technology transfer, from the pressure on developing countries and so on. The total overall picture is of a general system that appears to be aimed at goals other than those formally proclaimed. It is also a regime that attempts are being made to impose universally as the 'non-proliferation regime'.

An important number of developing countries wanting to provide their populations with the benefits of peaceful applications of nuclear energy have had to formulate their internal and external nuclear policies within this context. Given this situation and taking into account its consequences, some countries in the international community, and particularly some in Latin America, decided to pursue the autonomous development of nuclear energy for peaceful purposes and, because they had no alternative, to remain outside the NPT/IAEA system in order to retain their capacity for that development. It is significant that two of those countries in the region are the ones that have attained the highest levels of scientific and technological development in the nuclear field: Brazil and Argentina. Those two countries are also the ones willing to work toward international consensus on this subject, but under different principles. The degree of mutual confidence they have achieved can hardly be compared, in its practical results, with that under any other international control mechanism elsewhere.

A FUTURE OF CONFIDENCE ON THE BASIS OF COOPERATION

I have described, on the one hand, the special relationship between Argentina and Brazil in the field of peaceful use of nuclear energy and, on the other, some considerations about the so-called 'non-proliferation regime' and its objectives. When I analyse the NPT/full-scope safeguards/London Club system, I see an effort to prevent the proliferation of weapons by means of mechanisms that directly and negatively affect those countries wanting to develop. From a methodological viewpoint the so-called 'international non-proliferation regime' chooses to discriminate selectively against manifestations of nuclear development that it interprets as problems and frames the issue in terms of the acceptance or non-acceptance of certain control mechanisms. In contrast, the Argentine-Brazilian approach involves a comprehensive bilateral relationship based on and stimulated by mutual cooperation and confidence. This process, attained by means of a broad, solid and creative mechanism, does not foster political competition between the two states, with resulting beneficial consequences for regional and global security.

It is hard to imagine how the Argentine-Brazilian system may affect the NPT/IAEA/London Club system, since those countries have taken a firm stand against horizontal and vertical proliferation of nuclear weapons and have demonstrated more sensibleness in their behaviour than those proclaiming the monopoly of reasonability. The possibility that that regime could profit from the principles on which the Argentine-Brazilian relationship is based so as to overcome its weaknesses and inequities must be questioned.

From a purely formal point of view eventual coordination of the Argentine-Brazilian system and the NPT/IAEA system poses problems of incompatibility, given that the former measures confidence by a different unit and therefore does not entail a 'safeguards' regime (as understood by the IAEA) as is contemplated by Article III of the NPT and as transpired with the EURATOM treaty. However, if I compare the two systems, taking into account the impact of the Argentine-Brazilian system with respect to the formal goals of the NPT/IAEA regime, I realize that our system has by far exceeded, at a regional level, the expectations of the NPT/IAEA regime, since it thoroughly addresses the root of possible competition and simultaneously promotes cooperation and complementary actions that will undeniably yield future benefits to the peoples of both countries. These results are far more substantial, based on the goals originally established, than those obtained through the non-proliferation instruments themselves.

The two approaches are different, and the corresponding mechanisms have their own dynamics. When considering the Argentine-Brazilian system in terms of an eventual relationship with the NPT/IAEA system, the issue that emerges is nuclear energy usage as relates to safeguards, a focus that just touches the surface of the issue. The above does not mean that measures the two countries may agree to in the future may not become instrumentally compatible with the NPT/IAEA system.

Even if, hypothetically, the relationship between Brazil and Argentina moves toward the formulation of a bilateral regime – or eventually a multilateral one should other countries in the region become involved – that qualifies as safeguards, the possibility of co-ordinating with the NPT/IAEA system, given all the above issues, seems unlikely. The limited flexibility shown by the IAEA in the case of the safeguards negotiations concerning the Tlatelolco Treaty and the difficulty the European countries had in their EURATOM negotiations are the basis for that conclusion.

In light of the above considerations an attempt to project how the Argentine-Brazilian system will evolve on the basis of past experience is both forced and risky at present. However, to move somewhat in that direction, I might opine that the plans visualized for the bilateral relationship do not indicate any interest in establishing a mechanism that may complement the NPT regime.

As to the eventual application of the Argentine-Brazilian model else-where in the world, the particular circumstances of South America and, especially the subcontinent, have allowed for progress on integration pro-jects and the establishment of special confidence frameworks. A complex and heterogeneous set of technological, political and economic elements has made it possible for Argentina and Brazil to establish a system with unique features, extent and prospects. An attempt to apply this model to other regions implies trying to reproduce those particular circumstances in other parts of the world, an unlikely event. Every region has its own dynamics. However, world-wide acceptance of our experience may stimulate similar processes in other regions that may overcome past, rigid structures.

In any event, analysis of the Argentine-Brazilian nuclear development process provides some ideas on how to handle the application of nuclear technology at an international level that need to be noted:

1. If we actually want to respond seriously to the complex problems involved in this subject, we must expand the inventory of solutions proposed by the present system. The latter has reached its limits and is clearly insufficient for comprehensively addressing all the issues arising from the use of nuclear technology. If we do not want reality to go one way and the formal instruments of the so-called non-proliferation regime to go the other, we must overcome the conceptual limitations of the NPT/IAEA system.

2. As to the pursuit of non-proliferation on the terms of the NPT, its system is inadequate and has a negative impact on countries trying to pursue technological development and to create a capacity to determine their energy futures autonomously.

3. Analysis of this subject should not be limited to control mechanisms, to the disregard of the actual foundations of confidence. These comments are not aimed at judging the instrumental advantages of the present safeguards system as a control mechanism, since the latter constitutes an element to be taken into account and that must be enriched.

4. When we are faced with a political problem, we must look for a political solution. The instruments that will address the above problems will be the result of consensus that takes into account all positions and needs and that fundamentally address the post-war lack of confidence and confrontation. That process is the importance and originality of the Argentine-Brazilian system.

Presentation – *Counsellor José Felicio*

Before replying to the two questions addressed by this panel, we must first define what kind of bilateral arrangement we have in mind. Is it to be a system of nuclear controls or of nuclear cooperation?

Let me make clear at the outset that no system of controls regulates the nuclear cooperation between Argentina and Brazil. From the discussions we have had here, and from the discussions we have been having over the past four years at the Permanent Committee on Nuclear Policy, I see no intention on the part of either government to establish the kind of bilateral control regime suggested at this meeting.

If the need arises to bring certain aspects of our nuclear cooperation under controls, I am certain both governments will turn to the IAEA to ask for appropriate inspections. That may happen in the future. It has not happened so far. Therefore let me describe the nuclear cooperation arrangement that does exist today between Brazil and Argentina. I am pleased to have the opportunity to do so. In addition, I hope I will lay to rest any doubts that may still exist as to the peaceful intentions of the nuclear programmes of both countries.

The governments of Brazil and Argentina have always held identical positions with regard to the restrictions imposed by certain nuclear suppliers and to the discrimination that takes place under the NPT non-proliferation regime. The political views of the two states are similar. As for the technical aspects of their programmes, what Ambassador Roberto García Moritán has just said is correct. Brazil and Argentina chose different kinds of nuclear reactors for electricity production. Argentina's reactors use natural uranium for fuel, while Brazil chose the path of enriched uranium to fuel its nuclear power reactors.

This technical difference has not stood in the way of closer cooperation between the two countries in the nuclear field and has not impeded their ability to complement each other's programmes. For example, the two countries have always exchanged information and materials. This exchange increased after the agreement for nuclear cooperation, referred to by Ambassador García Moritán, was signed in 1980. For example, Argentina supplied Brazil with zircalloy tubes and uranium concentrate, and Brazil fabricated parts of the Atucha-2 reactor pressure vessel.

Nuclear cooperation was greatly intensified after Presidents Raul Alfonsín and José Sarney decided to promote the economic integration of Brazil and Argentina and signed the Declaration of Iguacu on 30 November 1985. On the same occasion they decided, in a joint declaration on nuclear policy, to establish a working group whose mandate would be to draw up the basis for bilateral nuclear cooperation. The group was and is still composed of members of the two ministries of foreign relations, atomic energy commissions and nuclear industries of the two countries.

Let me briefly cite some of the accomplishments of this working group, which later became the Permanent Committee. In July 1986 Protocol 11 to the Act of Integration was signed, establishing a basis for mutual assistance and exchanges of information in the event of a nuclear accident or radiological emergency. The world was then experiencing the impact of the accident at Chernobyl, and there was a clear need for greater international cooperation in the field of nuclear safety. Subsequently two international conventions with purposes similar to those of Protocol 11 were established within the framework of the IAEA.

Protocol 17 should also be mentioned. It created a political and technical framework for cooperation in such areas as reactor fuel elements, nuclear electronics and instrumentation, isotope enrichment, fast breeder reactors, safeguards techniques, nuclear and plasma physics and non-destructive assays. Protocol 17 is now being expanded to facilitate industrial cooperation and trade between the nuclear power programmes of Argentina and Brazil. As indicated, there will probably be greater participation by each nation's nuclear industry in the other's nuclear programme.

The advances in technical cooperation consistently have been followed by expressions of political support. Joint declarations on nuclear policy after Foz de Iguacu were signed in Brasilia, Buenos Aires, Viadema, Iperó and Ezeiza. These political declarations assured that bilateral cooperation could be extended to include other interested Latin American countries. It should be noted, too, that the Declaration of Iperó established the Permanent Committee on Nuclear Policy, to replace the working group created in Foz de Iguacu in 1985.

It was always understood that nuclear cooperation between Brazil and Argentina should continue beyond the Alfonsín and Sarney governments' terms of office. The idea has always been to strengthen cooperation and create mechanisms to ensure its continuation so that both societies may enjoy the benefits of the peaceful uses of nuclear energy.

Discussions in the Working Group and later in the Permanent Committee about establishing a bilateral control system never progressed beyond the talking stage. Cooperation developed so quickly that no control mechanism could effectively replace the intimate contacts that were taking place at the technical and political levels. It was clear that, with mutual confidence established through cooperation, there was no need for inspections.

Inspections, by the way, have a negative connotation. They denote suspicion, an undesirable aspect of any verification system. There is no place for such a mechanism in Brazil's bilateral cooperation arrangement with Argentina. If governments and experts from both countries think a bilateral control regime is unnecessary, then in whose interest would it be to create such an arrangement?

Let me now point to some facts that demonstrate the candour that exists between our nuclear authorities and the private sector. When the government of Brazil decided to announce in September 1987 that Brazilian scientists had mastered the technology of uranium enrichment using ultra-centrifuge machines, the first foreign government to be informed was the government of Argentina. President Alfonsín received a messenger sent to him by President Sarney for this purpose. When Argentina announced in 1983 that it had mastered the technology of uranium enrichment, Brazil was the first government to be informed.

There are other examples of this confidence. For example, President Sarney's visit to the Argentine uranium enrichment facilities at Pilcaniyeu was followed by the visits of Brazilian technical experts to these same installations. There have been visits to the excellent nuclear training centre Argentina has in Bariloche. The inauguration in April 1988 of the Aramar experimental centre jointly owned by the National Nuclear Energy Commission of Brazil and the Navy took place with the president of Argentina present. At that time Brazil's first uranium enrichment unit using the ultracentrifuge process was put into operation.

Brazilian experts visited a uranium reprocessing facility and heavy water production unit in Argentina. Our scientists are working together on a very interesting and important fast breeder reactor project. A joint delegation of experts from both countries is participating in an IAEA working group on breeder reactors.

I ask very frankly, Mr Chairman, what kind of inspection arrangement would replace this intimate arrangement? It is my opinion that a bilateral inspection mechanism between Brazil and Argentina would not serve to strengthen the IAEA's safeguards regime but would only satisfy the wishes of those who continue to advocate stricter and redundant controls over nuclear activities. If a bilateral inspection mechanism were outside the IAEA, it would probably affect the credibility of the IAEA. We should support the IAEA, not promote the creation of control mechanisms outside the IAEA framework.

As you shall see, we are continuing to work to promote close cooperation between Brazil and Argentina. There is no suspicion between these two nations. There is no bilateral inspection mechanism. We are not worried about bilateral inspections. What we have is cooperation. The best way to control nuclear proliferation, in my opinion, is to promote nuclear cooperation – not nuclear controls.

Response – *Dr Victor Gilinsky*

Could an arrangement for nuclear cooperation between Argentina and Brazil serve as a useful model for other regions? I am convinced, after a few days in your part of the world, that I should not be drawing parallels with other parts of the world. I will, however, give you my reaction to the nuclear cooperation arrangement that Minister Roberto García Moritán and Counsellor José Felicio outlined. The ultimate value of this cooperation is really for Argentina and Brazil to judge. There are many aspects that go beyond economic and technical considerations.

It is well to consider this arrangement in the context of nuclear power today. The growth of nuclear power for electrical generation has slowed in almost all countries and is even in retreat in some for a number of reasons. The technology has turned out to be a great deal more difficult and expensive than was forecast. However, the principal reason is safety. Safety has turned out to be more difficult to ensure than had been expected.

In view of this general contraction of nuclear power programmes, it makes sense for all countries, and certainly for Argentina and Brazil, to co-operate to avoid duplication of effort. It is especially useful to share operating experience with countries using similar technology. Whether or not their agreement meets these goals is ultimately for Argentina and Brazil to judge.

The same is true as far as the security aspects of the relationship are concerned. It is for you to decide whether informal arrangements are superior to formal inspections. While I detect in both presentations some possibility of detailed inspections in the future, for the time being both countries reject such inspection schemes, whether international or bilateral. In fact, I would say there is even a certain amount of hostility to the international inspection system and to its controls.

That position certainly affects other countries that may take the opinions of Argentina and Brazil into account when deciding whether to accept such international inspections. I have been trying to think how to explain how the reaction of other countries might be affected. An analogy was used so effectively yesterday by Counsellor Edmundo Fujita that I have tried to think of another one. Boarding the plane here from Buenos Aires we were delayed by the x-ray machine inspections, which leads me to the following thoughts. Everyone puts his hand luggage through the x-ray machine. You cannot excuse yourself by saying, 'I am a very reliable person.' No one seems to worry about it; no one thinks it is demeaning. Everyone just puts his handbag or briefcase through the machine. Let us suppose, on the other hand, that this procedure were voluntary. Suppose you were on a plane with about 100 passengers. Some passengers had their luggage x-rayed, some did not, and a couple of others said they had worked out a private arrangement and decided just to trust each other. Well, perhaps once in a while they would agree to look through each other's bags.

What would we think about this last arrangement? We would say it was better than none at all. It would make us feel somewhat safer. However, we would wonder why these passengers were not putting their briefcases through the machine. If they did, it certainly would be a lot easier to convince others who were holding out to join them and put their belongings through the machine, too.

Now, all analogies are imperfect. I know some of you thought ahead and noted that in the nuclear world of the Nuclear Non-Proliferation Treaty (NPT) there are some passengers who do not have to put all their baggage through the machine. That point is certainly true, but it does not help for everyone else to back away. I think that was the point that Mr Samuel Edlow was making earlier.

Nor are the various complaints that the inspection arrangement is too intrusive or that it is going to hamper this or that activity very convincing. In the real world, when we are talking about agreeing to 'intrusive' international inspections, we are talking about accepting what was accepted by the Federal Republic of Germany or Japan or Sweden or Canada or Australia. They are not bad company.

In fact, the problem with international inspections of the International Atomic Energy Agency (IAEA) is not that they are too strict. The problem is that they are too weak. We all need to think about that conclusion. Basically the whole system is too slow, in part for bureaucratic and in part for intrinsic reasons. To use the airport x-ray machine analogy, it would be as if the IAEA took a picture but developed it after the plane was in the air.

In a sense, that scenario is real. That is why it is dangerous to have unhampered use, world-wide, of all nuclear materials, especially the kinds that could be readily put to explosive use. I am talking about plutonium and highly enriched uranium, not about low enriched fuel and most of the material used in power production.

It is this concern that led the United States, almost thirteen years ago to the day, to change its policy in this area and to try to build more of a barrier between the use of nuclear materials for the generation of electricity and for military use. President Jimmy Carter gets most of the blame for this policy, but it was actually carried out by President Gerald R. Ford, shortly before the 1976 presidential election, on 28 October 1976. He called on other nations not to use plutonium for commercial purposes and at the same time, so that we would be doing what we asked of others, in effect barred its use in the United States.

As it turns out, plutonium use has become less and less economic. The reason is the reduced number of reactors, which consume less nuclear fuel overall than anticipated. Coupled with the discovery of lots of uranium, the price of uranium fuel has not gone up as anticipated. At the same time the capital costs of plutonium-fuelled reactors – fast breeders – have been going up. These facts make the whole thing look much less attractive.

Despite the unfavourable change in the economics of plutonium, there is a tremendous fascination in all research and development organizations with plutonium-based advanced nuclear technologies for a number of reasons.

It must be said that every country has a right to pursue its development as it sees fit. However, the point I would like to make is that if everyone pushes too hard and pushes to the legal limits of whatever sovereignty will permit, I think we will get ourselves into a lot of trouble on a worldwide basis. At Los Alamos there was a game during World War II in which the experimenter moved two subcritical masses of nuclear material forward with a screwdriver to see how close they could get without getting the experimenters into trouble. It was called 'tickling the dragon's tail'. Eventually someone got into serious trouble and died.

We all need to think seriously about the consequences of our pushing too hard on a world-wide basis using dangerous nuclear materials, which is, in effect, what we will be doing.

Discussion – *Ambassador Hector Gros Espiell*

In my opinion, the bilateral nuclear cooperation agreements between countries not party to the Nuclear Non-Proliferation Treaty (NPT) (such as the one between Argentina and Brazil) may advance the peaceful use of nuclear energy and the development of these countries. It is important to emphasize that they are an example worthy of analysis. While they may not be a model in the strict sense of the word, they can perhaps provide an example for countries under similar conditions in terms of technological and economic development and juridical framework.

These agreements have undoubtedly been based and must be founded – in a deliberate and juridically indisputable way – on the repudiation of all use of nuclear energy for purposes of war. For this reason I believe we have to find a formula for these agreements, or other similar ones that may arise in the future, which will have specific and appropriate regulations for international safeguards that will exclude all possible lack of trust and all possible susceptibility. I believe that no one has thought that agreements of this type must be beyond all type of control. However, we must attempt to find adequate controls, freely agreed on and not imposed. In my opinion international safeguards today do not have to be based on the NPT. However, that conclusion does not mean that the idea of international safeguards must be repudiated. The safeguards of the Treaty of Tlatelolco, which were rightly conceived as distinct from the NPT's, were freely discussed and agreed to in accordance with the particular characteristics of this treaty.

That treaty is not a discriminatory treaty as is the NPT. Unfortunately, however, as has occurred, international safeguards have almost always been a mere translation or projection of the NPT safeguards. They have not been negotiated with any flexibility. They have almost always been – sad to acknowledge – a form of contract of adhesion to the international level.

I believe that this situation should be overcome through an awareness on the part of Latin American countries, especially the countries that are party to the Treaty of Tlatelolco. In Latin America, for fundamental reasons of solidarity and cooperation, safeguards based on the Treaty of Tlatelolco for the countries party to the treaty should co-exist with other controls that provide trust, security and confidence for all Latin American countries not party to Tlatelolco. The current international situation leads us to believe that this widespread acceptance of safeguards will not be the case in the immediate future or the immediate short term. I believe that Latin American solidarity is an essential condition that must be saved at any price. This is an essential objective in itself, of interest to all Latin American countries whether or not they are party to the Treaty of Tlatelolco.

Now let me answer the question how can these cooperation agreements between Argentina and Brazil constitute a model for other areas? I said at the beginning of this brief presentation that they are not a model in a strict sense because they respond to the countries' own conditions and particular characteristics, which are not transferable. However, they can serve as an incentive for a quest for formulae similar to those that made regional cooperation and international trust possible. True universality – and I believe this point can be made today, although perhaps not five years ago – lies with the International Atomic Energy Agency (IAEA), a specialized agency of the United Nations, and no longer with the NPT. The IAEA should not be reduced to a mere agency for NPT application. Changes and evolution in the international situation compel us to consider what the NPT is today in relation to that reality and how today's changing reality will condition the future.

The important thing in the safeguard agreements, whether they be those of the IAEA, the NPT or the Treaty of Tlatelolco, is the objective they seek. They cannot be rigid and crystallized frameworks, nor can they be contracts of adhesion imposed on states. They must respect the distinct peculiarities of the countries and regions themselves. To impose a rigid and an unchangeable text is to provoke its violation by many who accept it in a merely verbal and formal way. The important things are the objectives and good faith, essential principles in international law for fulfilling what has been freely accepted.

The attitude of the Latin American countries party to the Treaty of Tlatelolco and of other countries not party to the agreement of cooperation between Argentina and Brazil, such as Chile, is fundamental respect for those cooperation agreements. The agreements are of interest to all of us, and no Latin American country can look at them from the perspective of an outsider. For example, it is very significant for Uruguay. Situated between Argentina and Brazil, it is party to both the NPT and the Treaty of Tlatelolco, while both its neighbours are party to neither the NPT nor the Treaty of Tlatelolco. Therefore Uruguay has a decisive interest in paying attention to and carrying out in-depth analysis and in pursuing nuclear cooperation between Argentina and Brazil. For that reason, these agreements have to be regarded, as I said at the beginning, as an effort worthy of the deepest interest and universal attention, but especially of the attention of Latin America. We must see that they are cooperation agreements for the peaceful use of nuclear energy. They are not essentially based on a repudiation *per se* of any international control, but only of those controls that are incompatible with its own nature.

Appendix: Nuclear Cooperation in the Context of the Programme for Argentine-Brazilian Integration and Cooperation

Monica Hirst and *Hector Eduardo Bocco*

THE NEW CONTEXT OF BILATERAL COOPERATION

The launching of the Argentine-Brazilian Integration and Cooperation Program (PICAB) in 1985 was the culmination of a process of bilateral *rapprochement* that had gained momentum since 1979. From the perspective of bilateral relations, three decisions mark this process: (1) the negotiation of the Corpus-Itaipu Agreement regarding the use of the Paraná River for hydroelectric power; (2) the understanding achieved during the Malvinas War, despite differing positions toward the Argentinean initiative; and (3) gradual formulation and implementation of the Process of Bilateral Integration announced at the end of 1985.

Indeed, solution of the essentially strategic conflict over the use of the hydroelectric resources of the Paraná River was decisive in settling old rivalries between the two countries. For Brazil the negotiation of the Itaipu-Corpus Agreement meant relinquishing a position of hegemony in relation to its neighbor, a position which it had held at different times over a period of 150 years. For Argentina it was important to have resolved a conflict that was incompatible with its aspirations for integration into the international economic and political system. Consequently, the hypothesis of bilateral conflict, always a part of the military doctrine of both countries, began to lose its strength.

Rapprochement broke new ground during the Malvinas War, a war which induced all Latin America to review the *de facto* 'rules of the game' of the inter-American system. For Argentina, this conflict revealed the limitations in the country's capacity for strategic initiatives and the

214

advantages of greater military cooperation with Brazil and growing diplomatic interdependence on strategic issues.

Finally, the deliberate effort by Argentina to correct the pervasive imbalances in bilateral trade relations laid the foundation for the Argentine-Brazilian integration programme. The success of this project depended fundamentally on its economic variable, which could not be successful in isolation. In addition to requiring the effective participation of the private and governmental economic sectors of both countries, it needed strong domestic political support in both countries, with important diplomatic backing on both sides, and without granting the Argentinean and Brazilian military sectors the power to veto.

It is noteworthy how this evolution marks a process of change in the premises that guide Argentine-Brazilian relations. The 1979 negotiations which sought to reconcile diverging interests implied an end to the typical competitiveness of a zero sum relationship. An intermediate situation was achieved during the Malvinas War. There was agreement on Argentina's right to the Malvinas, but Brazil opposed the use of force in territorial conflicts. At this time the two countries drew closer and took some common positions on decision-making and implementation (for example: Brazil represented Argentina's interests to Great Britain).

PICAB was established as a result of common decisions, at a time when an interest in cooperation was the prevailing spirit; the possibilities of agreement defined the content of the common bilateral agenda. Protocols signed during 1986–89 reveal an effort to place this project within a broader and more diverse context of bilateral links, encompassing political, military, economic, technological and cultural issues.

As in other integration processes, the principal incentive for the Argentine-Brazilian programme was political. Both governments thought they had a joint role to play in the region. Both believed that they should lead a process of definitive Latin American integration, thereby guaranteeing a dynamic place for the region in the international system. Thus foreign policies initiated to formulate and implement this Program sought to realize this joint role. Moreover, a combination of other factors justified the adoption of this role in the regional context: (1) the commitment of both governments to a process of transition to democracy; (2) the relatively greater development of the Argentinean and Brazilian economies and industrial structures; and (3) the determination of both countries to conduct autonomous foreign policies *vis-à-vis* the hegemonic power. There is a difference between the two countries' foreign policy objectives. Both gave great priority to bilateral integration. For Argentina, however, this had broader significance as one of the options for promoting the country's international reintegration, thereby representing a turn toward a new international praxis. Brazilian foreign policy has also been comprehensively redefined since the mid-1970's, and the fundamental premises of the new policy have remained in place since the initiation of civilian rule in 1985. Integration with Argentina should provide continuity to this process, with important short-term consequences for bilateral relations and medium and long-term consequences for all Latin America.

We can therefore conclude that PICAB is part of a long-term tendency in Argentine-Brazilian relations, but is also conditioned by the macroeconomic situation of both countries. Thus, after the harmony at the introduction of this initiative, the two countries moved apart to a certain extent. This was due, in particular, to external restrictions and to differences between policies aimed at redefining their participation in the world economy. This tendency toward integration is therefore marked by high and low points, and the topics on the agenda occupy primary or secondary roles accordingly.

At first the nuclear issue was unobtrusive on the common bilateral agenda, but it has come to play an increasingly central role as a result of the macroeconomic crisis of both countries. Finding common technological and military interests opened the possibility of consolidating nuclear cooperation, which after an unobtrusive beginning became the main political triumph in Argentine-Brazilian integration. The political impact of nuclear *rapprochement* between Argentina and Brazil largely counterbalanced the difficulties encountered in the economy. This situation also gave a new profile to the foreign policy agenda of both Argentina and Brazil. For the first time, this new area of cooperation actively connected the nuclear policies of the two countries and the positions they hold in the international arena.

BACKGROUND

Taking into account the historic competition between the two countries, and the regional supremacy and prestige attached to nuclear development, the bilateral *rapprochement* reached in 1980 represents a watershed. On this occasion, Argentina came to the negotiating table with a more favourable position than Brazil, by virtue of its general energy policies and its level of nuclear development. In contrast, Brazil was facing a significant drain of its foreign currency because of oil imports and had not yet been able to implement its nuclear plan (despite the contracts signed with West Germany), even though it had significant enterprises in hydroelectric power.[1] Nevertheless, the different technological directions taken by the two countries for their nuclear projects was the main reason for differences in the levels of accomplishment of each programme at the time.[2]

Argentinean dominance was confirmed by the fact that it was the only country in the region with an operating nuclear power plant (Atucha I) put into operation in 1974, and a significant, well-developed support infrastructure for uranium exploration, extraction and treatment, research reactors, human resources, radiological protection programmes, research programmes on the applications of nuclear energy, and so on. These achievements were possible because of Argentina's decision to develop nuclear power plants using natural uranium, thus allowing the country to act with greater autonomy.[3]

Argentina chose natural uranium-heavy water reactors, instead of other options (enriched uranium-light water), because it had access to the necessary technology and it cost less. Another reason was that the world supply of enriched uranium at the time was monopolized by the United States.[4] The military regime had grown distant from the US because of the Carter Administration's human rights and non-proliferation policies.

Having made this choice, the next objective was to control the entire nuclear fuel cycle, which meant gaining access to uranium enrichment technology, fuel fabrication, manufacture of zircalloy tubes, and production of heavy water.

Even though Argentina's choice of natural uranium/heavy water technology made complete control of the fuel cycle unnecessary, this decision was aimed at achieving energy autonomy and access to fast breeder reactor technology.[5] This objective gave rise to the Pilcaniyeu enrichment plant project whose positive results were made known in mid-1983.[6] Although one reason for this project was the possibility of a future shortage of natural nuclear material, we consider the main reason to have been the quest for access to the research reactor market, which had proved to be highly profitable at the time. The reactors sold to Peru and Algeria, and the expression of interest from a number of other countries in this type of technology, made the commercial prospects of nuclear development programmes of primary significance.[7]

The Argentinean government originally thought of this project as one in which other Latin American countries, especially Brazil, could participate. The unwillingness of Brazil, which stood by its decision to use enriched uranium and which did not want to subject its supply to changes in regional policies, resulted in this project being promoted by Argentina only.[8]

It is noteworthy that the level of development of Brazil's nuclear programme[9] in 1975 was far behind Argentina's programme. Brazil was still experimenting with trial generators and Angra I (a reactor bought from U.S. Westinghouse) was in the first stages of construction. Until 1975, the Brazilian programme depended entirely on US cooperation, as a result of Brazil's decision to use enriched uranium and light water technology.

This option, consistent with the decision to obtain the most advanced technologies,[10] thus making up for the differences with the Argentinean programme, encountered its first difficulties with the reluctance of the US to transfer sensitive nuclear technology. Also, the consequences of the 1974 Indian nuclear explosion and hardening of the US non-proliferation policy by the Carter Administration made themselves felt.[11]

When the relationship with the US showed clear signs of deterioration, relations with West Germany were intensified and resulted in the signing of the Agreements of 27 June 1975. Although this meant an important qualitative jump for Brazilian nuclear development, it caused considerable friction between each of these countries and the US and affected the type of cooperation the US adopted in its inter-bloc relations.

It is worthy of mention that during the same period, the Argentinean government unsuccessfully offered a similar agreement to Brazil. The Brazilian government showed little interest because it was about to sign its agreements with the Federal Republic of Germany,[12] and because of the adversarial relationship between the Brazilian military government and Argentinean Peronista rule.

When the Argentinean military took over the government (1976), a feeling of solidarity developed between the two countries in the face of pressures from the United States concerning the German-Brazilian agreements. For its part, the Argentinean government could foresee future conflicts with Canada – Argentina's principal nuclear supplier at that time – if that country backed the US non-proliferation position.

Simultaneously with its official programme, Brazil undertook a vast parallel research programme that intensified as implementation of the 1975 agreements slowed. After Argentina announced in 1983 that it had produced enriched uranium, the parallel programme became more active and embraced the following objectives: uranium enrichment by the gas centrifuge method[13] and by lasers; development of a compact nuclear reactor capable of propelling submarines and a fast breeder reactor. These objectives could match or complement the achievements of the Argentinean nuclear programme, and thus could provide the basis for collaboration as Argentina and Brazil acquired knowledge of complementary sensitive technologies. There is an interesting coincidence between the initiation of a parallel programme (1979) and the decision to move ahead in cooperation with Argentina in 1980.[14]

In summary, the nuclear policies of Argentina and Brazil at the time they signed the Agreements of 1980 were characterized by: the pursuit of different nuclear technologies; conflicts between Brazil and its suppliers; a firm decision by both countries to move ahead with their nuclear programmes; disassociation by both countries from the rule of international non-proliferation under the NPT;[15] refusal to adhere to the rules of the Treaty of Tlatelolco (together with Chile and Cuba); continuing financial problems in their respective programmes; and a greater consensus in Argentina than in Brazil – both at the political level and at the level of public opinion – with respect to nuclear development.[16]

The Agreements of 1980

An analysis of the relationships between the political regimes of both countries and the nuclear issues would merit a separate chapter. In both countries, the basic premise that sustains the nuclear programme – and for which a consensus exists regardless of the domestic debates – is technological autonomy, considered necessary for national sovereignty.

Two related subjects are especially relevant here. The first is symbolic and combines 'nationalism', military power, and autonomous science policies, all of which are characteristic features of the development models of both countries. The second is the question of military autonomy in the respective political regimes. This situation has continued to exist after the recent transitions to democracy in Argentina and Brazil, and hinders political control and openness toward public opinion regarding such crucial issues as the ultimate use of nuclear energy. Military interference in the nuclear field is known to be more of a problem in Brazil, where it is not easy to establish clearly the functional boundaries of the different institutions and companies involved in the programme.

The reasons that motivated Brazil to seek a *rapprochement* with Argentina on this subject include: its lagging nuclear programme and delays in the implementation of the 1975 agreement with the Federal Republic of Germany; the possibility (in spite of technological differences) of supplying Argentina with some components for the Embalse power plant and thus mobilizing the lethargic Brazilian industry, which by then had experienced shutdowns because of delays in Brazil's nuclear programme; the possibility of joint production of 250 and 300 MWe modular nuclear reactors – whose manufacture had been abandoned by the more advanced nations – that suit the energy needs and financial capabilities of the developing nations.

The new agreements of cooperation signed on 17 May 1980, in Buenos Aires, by Presidents Figueiredo (Brazil) and Videla (Argentina), provided for an exchange of technicians, training of personnel, and the exchange of information on the manufacture of components, physical protection of nuclear material, exploration for uranium, nuclear safeguards and reactor design research. In addition, Argentina would have access to the Brazilian Computerized Information Centre, would supply zirconium to Brazil, and would receive a supply of Brazilian enriched uranium for some Argentinean research reactors. It was further agreed that NUCLEP of Brazil was to construct part of the pressure vessel for the third Argentinean reactor supplied by the Federal Republic of Germany.[17]

The signing of these documents, as could be foreseen, produced a notable increase in nuclear relations between the two countries. These relations were characterized first by an emphasis on their political effect on the international community rather than the realization of effective scientific and technological cooperation, and second, by the establishment of balanced trade commitments that responded to mutual economic interests but that were not truly joint enterprises.[18]

The Malvinas War (1982) helped renew Brazilian misgivings about the ultimate intentions of the Argentinean military regime and delayed even further the implementation of the signed agreements. As of 1983, the particular internal political and economic situations contributed to the decline of Argentine-Brazilian nuclear cooperation.

Two conclusions can be drawn from this first stage of *rapprochement*. First, there were few concrete results with respect to specific cooperation, but these were of great importance considering the traditional rivalry between the two countries. In international terms, the consolidation of solidarity in multilateral forums served to reduce regional tensions and promote nuclear development in the face of international restrictions by means of reciprocal supplies and technical cooperation.

The second conclusion allows us to point to situations in which the process of cooperation, especially the highly sensitive area of nuclear cooperation, became vulnerable to a loss of trust in the attitudes of the other party. In the Malvinas conflict, despite their diplomatic solidarity, Brazil perceived Argentina's attitude as unilateral, opaque, and susceptible to being considered regionally destabilizing. One should recall that the main argument used by the nuclear powers to justify policies of selective non-proliferation specifically refers to the instability and unpredictability of political regimes such as those of Argentina and Brazil in the recent past.

The concept of an 'untrustworthy' country requires another, equally important definition. This refers to a supplier's capacity to provide for the needs of potential markets. In this respect, Argentina has developed an ability to respond to such requirements which rivals that of the members of the London Club as a trustworthy supplier, both because of Argentina's greater openness to technology transfers and because of its demonstrated responsibility in complying with signed agreements, limited only by its domestic economic circumstances.

Nuclear Cooperation Since the Transition to Democratic Governments

At the end of 1985, the foundations were laid for resuming bilateral *rapprochement* on the nuclear issue, as can be seen in the Joint Declaration on Nuclear Policy of Puerto Iguazú-Foz de Iguazú. This declaration marks the beginning of a new phase in nuclear cooperation, characterized by a series of political acts, visits by presidents and technical teams, joint declarations, signing of Protocols related to the PICAB, and establishment of work groups and periodical meetings.

President Sarney's visits to the Argentinean uranium enrichment plant in Pilcaniyeu (1987) and Ezeiza Nuclear Plant (1988), and President Alfonsín's visit to the Brazilian plant at Aramar (1988), represent the high point politically in the willingness of their respective nuclear programmes to 'open' their most sensitive installations to reciprocal inspection. Questions regarding the amount of enriched uranium in, and the production capacity of, each country have given rise to many suspicions, despite, in the Argentinean case, numerous official declarations.[19] These suspicions reveal the sensitivity of the subject for both countries and demonstrate that cooperation does not necessarily mean an absence of differences. Progress has been made by means of joint declarations, but no agreement yet has touched such issues as mutual inspections.

The Joint Declarations were distributed at the IAEA meetings, and given maximum publicity in order to show what was being done, the peaceful objectives of the respective programmes, and the cooperation between the two countries.[20] Between November 1985 and December 1989, there have been seven meetings of the Task Force (made a Permanent Committee in April, 1988), divided into groups for technical cooperation, coordination of foreign policies, legal and technical aspects, and so on. The purpose of the three principal binational subgroups is to: (a) unify international positions; (b) develop nuclear cooperation in the scientific-technical areas; and (c) develop proposals to establish systems of mutual safeguards. While the first two subgroups have been able to make progress, the third, entrusted with the politically difficult task of designing mechanisms and procedures that would bring about mutual openness in nuclear developments, has met with serious resistance from the military sectors of both countries. The

Argentinean and Brazilian projects for developing nuclear submarines – which are more advanced in the case of Brazil – and their achievements in long and short range rockets, make cooperation in these areas difficult and necessary.[21]

The more specific aspects of cooperation were defined in Protocols 11 and 17 of the Argentine-Brazilian Programme of Cooperation and Integration initiated in 1986. All meetings mentioned were held at the level of Secretary of State and Presidents of the Atomic Energy Commissions within the framework of Protocol 17.

Protocol 11 on Immediate Notification and Reciprocal Assistance in Case of Nuclear Accidents and Radiological Emergencies was signed in Buenos Aires, on 30 July 1986, anticipating the international conventions on 'Immediate Notification of Nuclear Accidents', and on 'Mutual Assistance in Case of Nuclear Accident or Radiological Emergency'. These were negotiated in the IAEA in September, 1986, as a direct consequence of the Chernobyl Nuclear Plant accident in the Soviet Union, and were put into effect only at the end of 1987. Protocol 11 received its 'baptism' with the Goiania accident, emphasizing that the international conventions signed after Chernobyl consider it a model on the subject.

Protocol 17, regarding nuclear cooperation, signed in Brasilia on 10 December 1986, details initial areas of cooperation in which examples of progress are already well-known, despite the budgetary problems of both countries. Most notable are: development of low enriched fuel for research reactors; exchange and development in nuclear instrumentation; research in nuclear fusion; cooperation in the implementation of IAEA safeguards; establishment of ten joint projects (already underway) in the areas of nuclear security and radiological protection, with a view to protection of the environment and preservation of health, and the establishment of a joint project for the development of technologies necessary for the construction of the future generation of nuclear reactors – fast breeders. This is certainly the most important enterprise, and it is estimated that a prototype will be built before the year 2010.

Businesses have taken important steps forward. An Argentine-Brazilian Enterprise Committee for nuclear issues has been formed and is working toward signing a Trade Protocol to guarantee equal conditions of participation for the respective national industries in competitive bidding for the Atucha (Argentina) and Angra II (Brazil) plants. This is very important because it is difficult to obtain international financing in the nuclear area for political and/or ecological reasons. Consequently one has to rely on one's own resources, which are scarce given the economic crises of both economies.

The few exchanges in the negotiating teams of each country – especially on the technical level – made possible greater *rapprochement*, ironing out differences and mistrust, and greater openness in the implementation of cooperation. The exchange of visits of technicians and scientists to each country's nuclear plants and experimental centres are particularly relevant (despite the fact that the visits of the presidents to the respective 'sensitive' centres were more valuable as spectacle).

Furthermore, the progress by the Permanent Task Force subgroup on technical cooperation has generated an appropriate framework for enhancing the scientific and technical teams' interest in learning about the accomplishments of their counterparts and in exchanging information. One cannot overlook the differences that exist within each scientific community with respect to the nuclear issue and the policies implemented by each government. In contrast to Argentina where there is a broader consensus, Brazil is engaged in an in-depth debate on the nuclear issue, and from the beginning has significantly questioned the decisions adopted during its process of nuclear development.

It is still difficult to evaluate precisely the technical achievements because: (a) progress in nuclear development has to be measured in the long term; (b) the economic crisis of both countries affects the scale of priorities in policy-making on the nuclear issue. In this context, the case of Argentina is exemplary considering the many budgetary constraints on the National Commission of Atomic Energy since 1983, in relation to the ambitious programme developed by the military regime in 1979.[22]

Formation of a Joint Agenda at Multilateral Forums

Although there is a long history of cooperation between Argentina and Brazil on the nuclear issue in international forums, the process of *rapprochement* that began in 1980 and gained momentum with the respective transitions to democracy has resulted in the formation of a new agenda of common initiatives.

The formation of this type of shared nuclear diplomacy has antecedents in the positions which each country has taken individually in multilateral forums. In the case of Argentina, its position can be expressed in the following way: to play an important role in all efforts to strengthen peace through negotiated settlement of conflicts and differences, and to promote international agreements that restrain or reverse the arms race, adopting three basic criteria. These are: (a) to give priority to containing vertical proliferation of nuclear arms; (b) to avoid horizontal proliferation on a universal and non-discriminatory basis, thus ensuring a balance of obligations between the States which possess and those which do not possess atomic weapons; and (c) to adopt measures of equal and balanced disarmament so that no state can claim advantage over any other at any stage.[23]

The main objective of Brazilian diplomacy with regard to nuclear proliferation and nuclear disarmament has been to maximize its contribution to world peace and security, by joining those nations that seek to end the decision-making monopoly of the states that control the technology and manufacture of non-conventional arms. With specific reference to the nuclear issue, Brazil's positions are based on two principles: domestically, to identify priorities in accordance with its national technological policy that places great importance on nuclear energy; in matters of foreign policy, to be guided by its international policy in defence of general and complete disarmament.[24]

As can be observed, the two positions coincide notably, thereby explaining the cooperation between the two nations in multilateral fora and in those relating to nuclear proliferation. The most notable antecedents in the formation of a common agenda are: (1) support for the Argentinean proposal presented in international fora regarding disarmament of chemical and bacteriological weapons. This was first expressed in 1970 and later reiterated, especially after 1983; (2) Argentinean support for the Brazilian initiative to convene the 'UNO Conference for the Promotion of Peaceful Use of Nuclear Energy'. This initiative was begun in 1980, but only recently realized after having overcome objections from the Nuclear Club regarding fear of proliferation; (3) Brazilian solidarity with Argentinean actions in the Group of Six; (4) Argentinean support of the Brazilian proposal to create a Peace and Cooperation Zone in the South Atlantic.

This convergence on nuclear and disarmament issues bore its first fruit with the request by both governments to the IAEA to participate as observers at the IAEA meetings on fast breeder reactors, not individually, but as one single observer. This initiative was taken in September 1988 and was well received by the IAEA, impressing and surprising its members.

The formation of a common agenda also relates to other international security issues, such as the conflict in southern Africa; the non-resolution of the Malvinas conflict; the future of the Antarctic Treaty; and the exploitation of South Atlantic mineral and fishing resources. Regarding the nuclear issue, some industrialized countries had hoped to use the sea-beds in the South Atlantic for disposal of toxic nuclear waste.[25] There was also a possibility of more nuclear submarines in the region, if as a consequence of the agreements between Washington and Moscow to reduce the number of missiles in Europe, there would be a corresponding increase in the number of nuclear submarines in the ocean depths and off the coasts of Africa and South America.[26]

Ecological issues have also been a concern, especially for Brazil because of pressures relating to the Amazon region. The Brazilian government opposed an attempt by some industrialized countries to link the environmental issue to nuclear proliferation at the meeting of the IAEA Governors (Vienna, February 1989). It argued that this was unjustified because there was increasing IAEA control of nuclear activities, and Brazil defended the principle of sustained development (thesis of the Brundlant-UN Commission) put forward by Brazilian diplomats, and meaning *development with environmental protection.*[27]

Progress in *rapprochement* and cooperation between the two countries has caused concern among the members of the Nuclear Club – especially the United States – but has not yet brought an increase in criticism, mistrust or pressures. Furthermore, some specialists – considered critics of the Argentinean and Brazilian positions on nuclear issues – contend that this confluence may facilitate the signing of the NPT, or at least, the formulation of a system of regional safeguards that approximate complete ratification of the Treaty of Tlatelolco.[28]

CURRENT STATUS OF ARGENTINE-BRAZILIAN NUCLEAR COOPERATION

The Nuclear Market

The nuclear issue, which by its very nature requires a high level of long-term planning and financing, has been affected by the crisis both economies have experienced and by the need to give priority to the demands of the short-term economic situation. In the case of Argentina, the harsh reality of this situation was seen in the collapse of the Energy Park constructed in mid-1988. Nonetheless, both countries have continued their most important enterprises, with some limitations, and Argentina fulfilled its commitments to projects in Peru (Huarangal) and Algeria (Nur), both now in operation.

These projects – considered unique because of the horizontal transfer of technology and the cooperation between two developing countries – are a model for demonstrating how to benefit from mutual progress by joining forces in the international market in a way that goes beyond shared nuclear diplomacy. This means exploring trade possibilities as a way to overcome the financial strangulation felt by every country in the nuclear field.[29]

The experience gained from this kind of enterprise can make technological autonomy feasible and beneficial. This can be explored in Argentine-Brazilian cooperation; both parties can contribute their achievements to pursue joint enterprises. Argentina has attempted to learn from the Brazilian experience in computer science and minicomputers, and from Brazil's ability to export technology that it has developed in recent years. The paradigm of technological autonomy can take place in the nuclear area by applying the strategies of shared nuclear diplomacy to joint enterprises.

Balance and Perspectives

In evaluating the nuclear programme, the cost-benefit relationship can only be viewed over the long term, given the nature of the nuclear enterprise. In the same way, more time is needed to evaluate the outcome of Argentine-Brazilian nuclear cooperation. Nonetheless, one may note the characteristics and complications of this process of *rapprochement*, and consider how these may influence its future course.

(1) First, the regional impact of the process is important because of the role the Argentine-Brazilian relationship plays in the propensity toward regional conflict or cooperation The cooperation that these two countries initiated and strengthened since their transition to democracy has helped to reduce their mutual suspicion and mistrust, thereby enhancing their *ability to lead* other players in the regional scenario. A clear sign of the positive effects of bilateral *rapprochement* on nuclear issues is evident in the attitude of OPANAL, which congratulated the Presidents of both countries for the steps taken to vindicate the use of nuclear energy for peaceful ends.[30]

(2) The democratic regimes of both countries have generated greater openness toward nuclear policy and made progress toward greater civilian control. Examples include President Sarney's decision to modify the Brazilian nuclear programme, the establishment of the Division of Nuclear Policies in the Argentinean Foreign Ministry, and the appointment of a civilian to head the CNEA. Likewise, the visits of the Presidents and technicians to the 'sensitive' plants of both countries has encouraged this openness.

Nonetheless, the tension that results from the degree of autonomy enjoyed by the Brazilian military sectors cannot be concealed, nor can the general phenomenon of military autonomy that can be observed to different degrees in both countries, and that brings with it an ability to exert pressure on political decisions.[31] Thus, the future of integration and cooperation are inevitably linked to the prospect of strengthening and deepening the democratic regimes in both countries.

Progress of mutual perception is evident from three events organized by the Joint General Staff of the Armed Forces of both countries[32], in which representatives of each nation met to exchange and debate military and strategic aspects of bilateral integration and cooperation. However, the presentation and discussion did not go beyond an exchange of ideas and statement of differences. The nuclear issue was not dealt with specifically, and was only mentioned, with no in-depth analysis. It may be a positive sign that the nuclear issue was not part of the military agenda, but since other issues that pertain to civilian decision-making were discussed – drug traffic, national security, and so on – the fact that it was not dealt with is significant.

(3) The foreign policy of both countries has made headway by establishing the nuclear issue as an area of specialization in the Foreign Ministry – Argentina established a Nuclear Policies and Disarmament Division in 1985 – and bringing together diplomacy with questions of technological development. This issue, which was almost totally neglected in the past by both countries, now plays a prominent role as technological issues have become areas of conflict in international relations. In addition to the specific problem of nuclear development, these include computer science, patents, pharmaceutical products, and fine chemistry.

(4) Internationally, Argentine-Brazilian cooperation has overcome a few initial misgivings, and has made a good impression at the IAEA. Although there are no explicit signs from the Nuclear Club with regard to the process of *rapprochement*, pressures continue to be exerted under the guise of other issues. For example: (a) Brazil's difficulties in obtaining a loan of US$500 million from the World Bank for electricity production, due to both ecological questions and to feasibility studies critical of the Angra III power plant; (b) the attempt to link the issue of environmental protection to the proliferation of nuclear weapons at the IAEA Governors' Meeting (February 1989); and (c) the traditional appeals to adhere to NPT rules.

One can see from these examples that pressure will continue to be applied, regardless of cooperation between the two countries – and even as a result of it – because what is in question is the issue of power linked to the paradigm of technological autonomy.

(5) In an attempt to make their peaceful intentions explicit, both countries supported the proposal for establishing a Peace and Cooperation Zone in the South Atlantic, in order to reduce tensions in the area[33] (in the context of the visible convergences in both countries' foreign policy agendas). One could see this decision as another attempt by both governments to establish

a system free of nuclear weapons and to eliminate once and for all the international mistrust as to the ultimate goals of their independent nuclear programmes.

(6) If one compares Argentine-Brazilian nuclear cooperation with the NPT system of non-proliferation, it becomes clear that the legitimacy of the former will depend on its effectiveness. The NPT has not been able to stop proliferation regardless of – or because of – the Treaty's rigid structure; therefore, the nuclear cooperation agreements between Argentina and Brazil can demonstrate – if their effectiveness is confirmed – that the pressures for non-proliferation are political as well as industrial. Behind the principles espoused by the Nuclear Club, one sees its intention to consolidate its monopoly over the nuclear industry.

(7) In this context, one may wonder why Brasilia and Buenos Aires find it difficult to make progress on the issue of mutual safeguards and controls, as these are necessary to sustain the positions of both countries. No explanations are given in official statements; when information is requested from officials the answer is more or less: the issue is discussed but its importance does not invalidate the debate of other equally important subjects on the bilateral agenda. This diplomatic language reveals the difficulty in making progress on the issue, and the lack of proposals that satisfy both parties.

The difficulties in reaching some kind of consensus seem to be rooted in both countries, but are more evident in Brazil,[34] since the Brazilian foreign ministry (Itamaraty) and the military sectors cannot agree on control mechanisms for the nuclear programme.

This is mainly due to the persistence of reciprocal mistrust whose causes include: imbalance in the level of nuclear development, different levels of technological autonomy, and suspicion between the military sectors. Moreover, similar situations, such as that of Pakistan in relation to India (which has not had a positive outcome), does not encourage the construction of a system of safeguards among non-central countries.[35]

Even though a proposal which satisfies both parties has not yet been developed, the political atmosphere around the process of nuclear cooperation has been highly favourable, making it difficult to imagine 'another context in which the negotiation of a mutual control mechanism would have better possibilities of success'.[36] Yet these difficulties have not slowed down other aspects of Argentine-Brazilian nuclear cooperation.

(8) Another complex subject is how far each country has progressed toward development of a nuclear-propelled submarine. Brazil has initiated the project, favoured by a large military involvement in the nuclear programme and considerable financial resources, given the greater budgetary autonomy of its armed forces, which is directly related to their level of

autonomy in the domestic political system.[37] But Argentinean achievements in uranium enrichment and the capability this provides for nuclear propulsion have been closely observed.

Paradoxically, both countries have better technological conditions to construct a military nuclear device than to produce a submarine with nuclear propulsion. The complexity and cost of the programme would be higher, apart from the difficulties that would be involved in manufacturing the first prototype.

Argentina and Brazil have extensive coastal areas and such a project would be tempting for both, but to what use each country would put it, and which country will be the first to manufacture it if it were a joint project – since the financial conditions to produce two prototypes would not exist – is still a question and implies a level of harmony and shared strategies which currently do not exist. Joint development therefore requires a level of debate and consensus yet to be developed.

The problems of nuclear cooperation between Argentina and Brazil are clearly very complex and can only be resolved at higher levels of understanding and cooperation than those reached by the current level of *rapprochement*. Nonetheless, significant political progress has been made, and the political framework that has been developed is optimum for moving to higher levels of convergence. Furthermore, the process of *rapprochement* and cooperation is not over, but constantly being improved, and currently nuclear cooperation can be considered as PICAB's main political triumph. Nuclear questions should be considered in the long term, as this will permit an evaluation of whether *rapprochement* will be able to resolve the complex difficulties implied in the current situation of nuclear cooperation between Argentina and Brazil.

Notes

Introduction

1. The report and papers of the Task Force were published as *Preventing Nuclear Terrorism*, Paul Leventhal and Yonah Alexander, editors, Lexington Books, 1987, 472 pages. The principal papers presented at the Buenos Aires meeting, evaluating the Task Force's findings, are available from the Nuclear Control Institute.
2. Under the conference groundrules, statements made during the discussion periods following the formal presentations and responses were not to be attributed to the speakers by name.

1 Rapporteur's Summary

1. Nuclear fuel cycle involves the enrichment of uranium, fabrication of fuel for nuclear reactors, reprocessing of spent fuel to separate the plutonium byproduct, and fabrication of fresh fuel from the plutonium and/or uranium.
2. Most experts view a peaceful nuclear explosion as indistinguishable from the test of a nuclear weapon. The NPT permits PNEs if they are conducted by a nuclear weapon state on behalf of a non-nuclear weapon state. In 1974 India, which has not ratified the NPT, demonstrated a nuclear weapons capability by setting off an underground PNE. The Tlatelolco Treaty permits PNEs, although some interpret the treaty as defining weapons in such a way that a PNE could not be tested without violating the prohibition of nuclear weapons.
3. Certain suppliers party to the NPT, such as the United States and Canada, require acceptance of full-scope safeguards as a condition for the supply of reactors and fuel – that is, safeguards on all nuclear facilities, both indigenous and imported.
4. Here nuclear submarines refer to submarines powered by nuclear reactors that do not carry nuclear weapons. The standard acronym for such boats is SSN.
5. Admiral Mario Cesar Flores, 'Submarino de Propulsao Nuclear', *Revista 'O Periscopio'*, XLIII (1988) 6–10.
6. The 'London Club' is the term applied to a group of advanced nuclear technology nations that entered into an agreement following a meeting in London in 1974 to limit nuclear exports.
7. The 1978 Nuclear Non-Proliferation Act (NNPA) prohibits exports

230

of nuclear reactors or fuel to nations, including Argentina, that have not signed the NPT or otherwise accepted full-scope international safeguards.

2 Goals of Argentine-Brazilian Nuclear Cooperation

1. The views in this paper are the author's and not necessarily those of the institution to which he belongs.
2. A. F. Montoro, 'Importancia Politica dos Acordos Argentina-Brasil' [Political Importance of the Argentine-Brazilian Agreements,] a paper presented at the III International Seminar 'Argentina-Brasil: Perspectivas Comparativas y Ejes de Integración' [Argentina-Brazil: Comparative Perspectives and Means of Integration], sponsored by the Facultad Latinoamericana de Ciencias Sociales (FLACSO), Buenos Aires, October 1988, p. 4.
3. This statement was made in Rio de Janeiro on 24 August 1910, when Sáenz Pena was returning from Europe to assume the presidency of the republic. Isidoro Ruiz Moreno maintains, however, that the real author of this statement was General Julio A. Roca, twice president of the republic, who in 1907 said, while on a trip to San Pablo, 'Nothing divides us and everything brings us nearer'. In I. Ruiz Moreno, *Historia de las Relaciones Exteriores Argentinas* [History of Argentina's Foreign Relations] (Buenos Aires: Editorial Perrot, 1961) p. 91.
4. M. Hirst, 'Las Perspectivas del Diálogo Bilateral' [Perspectives on the Bilateral Dialogue], in M. Hirst (ed.), 'Argentina-Brasil. El Largo Camino de la Integración' [Argentina-Brazil. The Long Road to Integration] (Buenos Aires: Editorial Legasa, 1988) p. 193.
5. C. M. Muniz, *Las Relaciones entre la Argentina y Brasil* [Relations between Argentina and Brazil] (Buenos Aires: Museo Mitre, 1979) p. 45.
6. Statement made at a conference at the Rotary Club of Palermo, Buenos Aires, June 1969, cited by Muniz, *ibid.*, p. 71.
7. M. Hirst, 'El Programa de Integración Argentina-Brasil: De la Formulacion a la Implementacion' [The Program of Argentine-Brazilian Integration: From its Planning to Its Implementation], Serie Documentos e Informes de Investigacion No. 67, FLACSO, Buenos Aires, July 1988, p. 21.
8. C. Martinez Vidal-R. Ornstein, 'La Cooperación Argentina-Brasil en el Campo de los Usos Pacíficos de la Energia Nuclear' [Argentine-Brazilian Cooperation in the Peaceful Uses of Nuclear Energy], a paper presented at the seminar organised by FLACSO, see note 2, p. 2.
9. An extreme example of this unjustified attitude is the article by Richard Kessler, 'Peronists Seek Nuclear Greatness', *Bulletin of the Atomic Scientists* (May 1988) pp. 13–15.

10. An example of this view is found in the article by M. J. Culaciati, 'El Acuerdo Nuclear con Brasil' [The Nuclear Agreement with Brazil] *La Prensa*, 25 April 1988, in which it is written that 'Although the agreement with Brazil is a step in the right direction, it is insufficient if it is restricted to the small sphere of bilateralism and surely will fail . . . [to] avoid us still being suspected of using nuclear energy for war or war related purposes.'

11. A typical statement to this effect was that, for example, by Paul Leventhal, president of the Nuclear Control Institute of Washington, D.C., in *The New York Times*, 22 July 1987.

12. 'Newsbrief', *Programme for Promoting Nuclear Non-Proliferation*, No. 6 (July 1989) 2, published by the Centre for International Policy Studies, Department of Politics, University of Southampton, U.K., p. 2.

13. J. Goldblat, 'Nuclear Non-proliferation: The Status and Prospects', Background Paper No. 29, Canadian Institute for International Peace and Security, Ottawa, June 1989, p. 7.

14. Report of the rapporteur of the 55th Workshop on 'Non-Proliferation and the Non-Proliferation Treaty' organised by the Pugwash Conferences on Science and World Affairs, Dublin, 5–8 May 1989.

15. It is uncommon, for instance, that two countries are represented by the same person at an international meeting, as occurred with Argentina and Brazil at the April 1989 meeting of the Working Group of the IAEA on Breeder Reactors (*Nucleonics Week*, 6 April 1989).

3 Industrial and Economic Benefits of Latin American Nuclear Cooperation

1. Henning, Fernando A. S. *Nuclear Technology in Brazil.*
2. Bernal Castro, José. *Argentine Nuclear Development.*

4 Models for a Bilateral Confidence-Building Regime

1. United States Atomic Energy Act of 1954.
2. *IAEA Safeguards, Guidelines for States' Systems of Accounting for and Control of Nuclear Materials.* International Atomic Energy Agency, Vienna, 1980. IAEA/SG/INF/2.
3. *The Physical Protection of Nuclear Material.* International Atomic Energy Agency, Austria. June 1977. INFCIRC/225/Rev. 1.
4. Rozental, J. J., 'Physical Protection Philosophy and Techniques in Brazil'. *Journal of the Institute of Nuclear Materials Management.* January 1988.

5.   IAEA Safeguards are described in the series of IAEA publications: *IAEA Safeguards*, IAEA/SG/INF/1–6. International Atomic Energy Agency, Vienna. These publications are: *IAEA Safeguards Glossary*, 1987, IAEA/SG/INF/1 (Rev. 1); *IAEA Safeguards, Guidelines for States' Systems of Accounting for and Control of Nuclear Materials*, 1980, IAEA/SG/INF/2; *IAEA Safeguards, An Introduction*, 1981, IAEA/SG/INF/3; *IAEA Safeguards, Aims, Limitations, Achievements*, 1983, IAEA/SG/INF/4; *IAEA Safeguards, Safeguards Techniques and Equipment*, 1984, IAEA/SG/INF/5; *IAEA Safeguards, Implementation at Nuclear Fuel Cycle Facilities*, 1985, IAEA/SG/INF/6.

6.   *International Nuclear Fuel Cycle Evaluation, INFCE Summary Volume*, International Atomic Energy Agency, Vienna, 1980. INFCE/PC/2/9. ISBN 92–0–159980–3.

7.   David Fischer and Paul Szasz, *Safeguarding the Atom: A Critical Appraisal*, SIPRI (London: Taylor & Francis, 1985) p. 131.

8.   Hexapartite Safeguards Project, J. Menzel (ed.), 'Safeguards Approach for Gas Centrifuge Type Enrichment Plants', *Nuclear Materials Management* (Winter 1983) 30-7.

9.   See George Bunn and John B. Rhinelander, 'Deep Cuts in a Peaceful World: Steps Toward a Minimum Deterrent After Start', a paper prepared for the 7th Annual Conference of Lawyers Alliance for Nuclear Arms Control and Association of Soviet Lawyers, Moscow, October 1989.

10. Don O. Stovall, 'A Participant's View of On-Site Inspections', *Parameters*, US Army War College (June 1989) 1–17.

11. Ibid.

12. Op. cit., Bunn and Rhinelander.

13. Robert Pear, *New York Times*, 18 July 1989, p. 1.

14. Thomas L. Friedman, *New York Times*, 24 September 1989, p. 16.

15. See Brigadier General Roland Lajoie, 'Insights of an On-site Inspector', *Arms Control Today* (November 1988) 3–10.

16. Through the end of May 1989 US and Soviet teams had conducted a total of 340 inspections at INF sites, 96 by Soviet inspectors in the United States and Western Europe and 244 by US inspectors in the Soviet Union and Eastern Europe. One hundred forty-six of these were baseline inspections carried out by the two sides in nine countries at all INF facilities, 115 at Soviet facilities and 31 at US facilities. By early July the United States had destroyed 325 missiles out of 846 missiles to be eliminated under the treaty while the Soviet Union had destroyed 1038 missiles of its 1 846 treaty-limited missiles. ('Nuclear Notebook', *Bulletin of Atomic Scientists* September 1989); *Washington Post*, 3 June 1989, p. 3.

17. Michael Krepon and Sidney N. Graybeal, *Arms Control Today* (November 1988) 10-14.

18. Op. cit., Lajoie.

19. Op. cit., Krepon and Graybeal.

20. R. Jeffrey Smith, *Washington Post*, 22 September 1989, p. A33.
21. Lewis A. Dunn and Amy E. Gordon, 'On-Site Inspection for Arms Control Verification: Pitfalls and Promise', Center for National Security Negotiations, McLean, VA, May 1989.
22. Johan Swahn, 'Open Skies for All: The Prospects for International Satellite Surveillance', Technical Peace Research Unit, Chalmers University of Technology, Goteborg, Sweden, January 1989, p. 5.
23. For a discussion of the uses of satellite imagery see ibid. and Ann M. Florini, 'The Opening Skies: Third-Party Imaging Satellites and U.S. Security', *International Security* (Fall 1988) 91–123.
24. Landsat is operated by the US-subsidised Earth Observation Satellite Company (EOSAT); the Belgian and Swiss governments have a small number of shares of the French-regulated company SPOT, Image. Some civilian imaging is now being supplied by the Soviet companies, Soyuzkharta and Glavcosmos, from photographic film, which is dropped from various Soviet military satellites.
25. Jeffrey Richelson, 'Military Intelligence – SPOT Is Not Enough', *Bulletin of Atomic Scientists* (September 1989) 26.
26. Peter Zimmerman, 'From the SPOT Files: Evidence of Spying', *Bulletin of Atomic Scientists* (September 1989) 24.
27. For a complete discussion see Leonard S. Spector, 'Monitoring Nuclear Proliferation', in *Commercial Observation Satellites and International Security* (New York: St. Martin's Press, 1990).
28. Ibid.
29. Ibid.
30. 'The Implications of Establishing an International Satellite Monitoring Agency', United Nations, New York, 1983, p. 6.
31. About one-third of the average 100 military satellites launched each year have been for photo-reconnaissance by the United States, the USSR and China. The Soviet Union launches many more reconnaissance satellites than does the United States because its satellites have much shorter lifetimes – fourteen days compared with three years. In 1986 China recovered its eighth photo-reconnaissance satellite. France, with support from Spain and Italy, has been developing the Helios military reconnaissance satellite. (See Bhupendra Jasani, 'Military Use of Outer Space', *SIPRI Yearbook 1987* (Oxford: Oxford University Press, 1987) p. 59.) The United States and the Soviet Union began to launch imaging satellites in 1962. These first satellites had short lifetimes of a month or less and used photographic film that required measures for ejection and recovery. Since then, satellite lifetimes have lengthened considerably, and the use of film has been replaced by electro-optical sensors that collect digitized information that is sent back immediately to receiving stations on earth. The US KH-11 satellites, first launched in 1976, have lifetimes of two to three years; Soviet satellites now have lifetimes ranging from 14 to 200 days. (See Bhupendra Jasani, op. cit., p. 60; Jeffrey Richelson, *The U.S. Intelligence Community* (Cambridge:

1985) Ballinger, pp. 107–115.; *Sword and Shield*, (Cambridge, 1986) Ballinger, pp. 90–7.) The advanced US KH-12 satellite with area surveillance, high-resolution close-up and radar imaging capabilities was reported to have been launched in August from the Columbia space shuttle. (William Broad, *New York Times*, 9 August 1989, p. A14.)

32. Op. cit., Swahn, p. 236.
33. Op. cit., Jasani.
34. Op. cit., Richelson, *Bulletin.*
35. Senate Armed Services Committee, FY1984 DOD Authorization, Part 6, pp. 2948, 2949.
36. Thomas Friedman, *New York Times*, 24 September 1989, p. 16.
37. Op. cit., Richelson.
38. Op. cit., Richelson, *U.S. Intelligence Community*, p. 116.
39. Private communication from John Pike, Federation of American Scientists.
40. Arthur F. Manfredi *et al.*, 'Ballistic Missile Proliferation Potential in the Third World', Congressional Research Service, Washington, DC, April 1986; Robert D. Shuey *et al.*, 'Missile Proliferation: Survey of Emerging Missile Forces', Congressional Research Service, Washington, DC, 3 October 1988, revised 9 February 1989.
41. Op. cit., Swahn, p. 69.
42. Statement by the Assistant to the President for Press Relations, 16 April 1987.
43. Michael Krepon, 'Peacemakers or Rent-a-Spies?' *Bulletin of Atomic Scientists* (September 1989) p. 12.
44. Military surveillance satellites traditionally have used film because of its high resolution, but they are making the transition to electro-optical devices of equal capability. Commercial imaging satellites use electro-optical devices, except for Soyuzkharta, which markets images that have been recorded on film. One disadvantage of film is that the effective life of the satellite is limited to the film supply. In addition, film must be processed, either on-board and transmitted back to earth by television or after the film has been ejected and collected, so that there is a delay in receiving the pictures for intelligence or verification purposes. These limitations often are not important for military satellites because they may fly in orbits so low (for example, 200 metres above the earth) that atmospheric drag limits lifetimes to weeks or months. Commercial satellites are expected to remain aloft for several years at heights on the order of 700 metres (for Landsat). The digitized information gathered by the sensors can be transmitted on-line to receiving stations around the world in line of sight of the satellite. (See also endnote 41.)
45. For a discussion of imaging technology and science see Kosta Tsipis *et al.* (eds), *Arms Control Verification* (New York: Pergamon-Brassey's, 1986) pp. 63–120; John R. Schott, 'Remote Sensing of the Earth: A

Synoptic View', *Physics Today* (September 1989), 72; and Richelson
Bulletin, Appendix C, op. cit.

46.   For the thematic sensor of Landsat 4 and 5 the pixel resolution or IFOV
       is 30 metres in each of six visible or near infra-red color bands and 120
       metres in a thermal infra-red band. For Spot I, the pixel resolution is
       10 metres in panchromatic (black and white) and 20 metres in three
       spectral bands. Military satellites may have IFOVs as small as several
       centimetres.

47.   The resolution of a satellite image improves with a smaller pixel, a
       larger camera focal length and a decreased satellite height. Other
       factors are the size and shape of the object, ground contrast, the
       light-gathering ability of the sensor optical system and system noise.
       For film, the ground resolution is measured by the distance between a
       pair of lines on the earth that can just be resolved on the film. Generally
       film resolution, which is a practical indicator of picture quality, is 2.0
       to 2.5 times the pixel resolution, depending on the other factors cited
       above.

Hexapartite Appendix Bibliography

'Safeguards Approach for Gas Centrifuge Type Enrichment Plants', *Journal of Nuclear Materials Management* (Winter 1983), authored by the
Hexapartite Safeguards Project Committee and edited by Joerg H. Menzel.
'Hexapartite Safeguards Project Overview', *ESARDA Bulletin* (October 1983).
The five papers in the section on *NDA in Centrifuge Enrichment Plants* in the
1987 ESARDA Proceedings of the 9th Annual Symposium on Safeguards
and Nuclear Materials Management (ESARDA 21).

5 Nuclear Submarines and their Implications for Weapons Proliferation

1.   A. Carnesale, 'Nuclear Power and Nuclear Proliferation', in Options for
      U.S. Energy Policy (San Francisco: Institute for Contemporary Studies,
      1977) p. 67.

2.   Section 127 of the Atomic Energy Act of 1954 as amended by Section
      305 of the US Nuclear Non-Proliferation Act of 1978.

3.   'Guidelines for Nuclear Transfers', agreed to in London on 11 Sep-
      tember 1977 by the Nuclear Suppliers Group, the so-called London
      Club, and attached to communications addressed on 11 January 1978
      to the Director General of the International Atomic Energy Agency
      (IAEA). It was published by the IAEA as INFCIRC/254, IAEA,
      Vienna, 1978, Article 7, 'Special controls on sensitive exports'. The
      London Club deliberations actually predated the Carter administration

– they began in 1975—and the guidelines were weaker than the Carter policy, especially in terms of not requiring full-scope safeguards as a condition for nuclear exports.

4.   Ibid., Article 8, 'Special controls on export of enrichment facilities, equipment and technology'.

5.   US Nuclear Non-Proliferation Act of 1978, Section 502.

6.   A. Wohlstetter *et al.*, Swords from Plowshares (Chicago: University of Chicago Press, 1979).

7.   See, for example, R. H. Williams and H. A. Feiveson, Diversion-Resistance Criteria for Future Nuclear Power, Report No. 239, Center for Energy and Environmental Studies, Princeton University, June 1989.

8.   In principle plutonium could also be used, but the radioactive hazards in fuel fabrication as well as in routine operation on board the submarine would be much higher.

9.   On 7–8 July 1989, during a trip arranged by the Natural Resources Defense Council (NRDC) and the Soviet Academy of Sciences, a group of Americans visited the Soviet nuclear materials production complex at Kyshtym. During the visit they were briefed by Soviet scientists. During one such briefing Evgeny I. Mikerin, head of the main department of manufacture and technology of the USSR State Committee for the Utilisation of Atomic Energy, told them that 'Soviet naval reactor uranium is about 10% uranium-235.' See Fact Sheets released by the NRDC, Washington, D.C., 11 July 1989.

10.   This point is especially true in the case of centrifuge technology.

11.   D. D. Lanning and T. Ippolito, 'Some Technical Aspects of the Use of Low-Enriched vs. High-Enriched Uranium Fuel in Submarine Reactors', presented at the Conference on the Implications of Acquisition of Nuclear-Powered Submarines (SSN) by Non-Nuclear Weapons States, MIT, Cambridge, Mass., 27–28 March 1989, to be published.

12.   Typically SSNs spend approximately 240 days per year at sea, and their reactors operate at about 25 per cent of full power on average. Thus the number of full-power days per year is about 60.

13.   For example, the spent fuel discharged annually from a 1,000 MWe PWR operating at a 75 per cent capacity factor contains about 230 kg of plutonium; the spent fuel discharged annually from a 1,000 MWe CANDU reactor at the same capacity factor contains about 490 kg of plutonium.

14.   This figure is in agreement with the plutonium production of the French low-enriched reactor given by Yves Girard, 'Development and Current Status of French SSN Propulsion Technology', MIT Conference, op. cit.

15.   Ibid.

16.   T. B. Cochran, W. M. Arkin, R. S. Norris, and M. M. Hoenig, Nuclear Weapons Databook, Volume II: U.S. Nuclear Warhead Production, (Cambridge, Mass.: Ballinger, 1987) p. 71.

17. For a comprehensive analysis of the safeguards issues in the NPT context see M. F. Desjardins and T. Rauf, Opening Pandora's Box? Nuclear-Powered Submarines and the Spread of Nuclear Weapons, Aurora Papers 8, The Canadian Centre for Arms Control and Disarmament, Ottawa, Canada, June 1988. See also B. Sanders and J. Simpson, Nuclear Submarines and Non-Proliferation: Cause for Concern, Occasional Paper Two, Programme for Promoting Nuclear Non-Proliferation; Centre for International Policy Studies, University of Southampton, Southampton, England, July 1988.
18. D. Fischer and P. Szasz in J. Goldblatt (ed.), Safeguarding the Atom (London and Philadelphia: Taylor and Francis, 1985) p. 81.
19. I thank Lawrence Scheinman for an interesting discussion on this point.
20. For a Brazilian naval perspective on this issue see 'The Nuclear-Powered Submarine', Brazilian Maritime Review, May 1988.
21. In particular, an SSN armed with conventional anti-ship cruise missiles such as the submarine Harpoon would be a formidable offensive or defensive threat. On the other hand, a diesel-powered submarine can also carry nuclear weapons on ballistic or cruise missiles, torpedoes or depth charges.
22. For a general discussion of the relative merits of the fuel cell, Stirling engine and nuclear hybrid alternatives see C. A. Prins and A. A. Ham, 'In Search of Air-Independence', Maritime Defense, March 1988, pp. 75–80.
23. The task of safeguards in such a fuel cycle is simplified by the fact that weapons-usable materials do not legitimately appear. Further, the non-use of such materials greatly reduces the risk of sub-national diversion and subsequent manufacture of a nuclear weapon.
24. M. Sakitt, Submarine Warfare in the Arctic: Option or Illusion (Stanford, Calif.: Center for International Security, Stanford University, 1988) pp. 57–63.
25. Revista 'O Periscopio', XLIII (1988) 6–10.

Appendix: Nuclear Cooperation in the Context of the Programme for Argentine/Brazilian Integration and Cooperation

1. We refer especially to the hydroelectric complex of Itaipu in Alto Paran, which also caused many conflicts with Argentina up to the signing of the 1979 agreements. In the Alto Uruguay: San Pedro, Garavi, and Roncador.
2. Pedro, Marta y Pouget, Mario G. Martin, El Plan Nuclear del Gobierno del Proceso de Reorganizacion Nacional y su proyeccion global, regional, y nacional, en Investigaciones de Politica Internacional, Univ. Nac. de Cuyo, Mendoza-Argentina, 1985.
3. Ibid.

4.   On ending the U.S. nuclear industrial monoply, see Joskom, Paul L., The International Nuclear Industry Today: the End of the American Monopoly, in Foreign Affairs, 54 (July 1976):789–803. The concern for non-proliferation may be related more to industrial monopoly than to atomic danger. A good example is the case of India's 1974 nuclear explosion, the principal assistance for which (in the form of a reactor used to produce plutonium) came from Canada, the United States' main partner in non-proliferation issues.

5.   It should be recalled that knowledge of uranium enrichment is important for manufacturing nuclear devices for military use, even though this method is very costly (the other method uses plutonium). This cost argument is used by Brazilians and Argentineans to allay fears regarding the nuclear development of both countries.

6.   Among other objectives, Argentina considered becoming an exporter of enriched uranium, competing with the United States and producing its own materials for research reactors. Nonetheless, we consider the dispute to be essentially related to nuclear power issues.

7.   Another argument in favour of achieving uranium enrichment capability was the refusal of the United States to provide enriched uranium for reactor that the Argentinean government had agreed to build for Peru.

8.   One should note that the gaseous diffusion method for enriching uranium used by Argentina requires large plants, unlike the ultracentrifuge method developed by Brazil in the Aramar Plant. If Argentina is to increase its enrichment capability, it will have to install large equipment easily detectable from the air, while Brazil currently has 300 parallel centrifuges and intends to install 3,000. With 1,000 cascades grouped in series Brazil is planning on enriching uranium to 90% very quickly and without large plants. See *El Cronista Comercial*, Buenos Aires, 8–12–88. These capabilities are already known; whether they can be observed by air or by satellite is irrelevant. Nonetheless, the type of plant to be built is not irrelevant.

9.   For a more detailed history, see: Grabendorf, Wolf, La politica nuclear y de no proliferation de Brasil, in *Estudios Internacionales*, Santiago de Chile, Ano XX, octubre–diciembre 1987, Nro. 80.

10.   A constant objective of Brazil's nuclear policy was to establish cooperation agreements with the most technologically advanced countries in order to gain access to technologies that were not available domestically and transfers of which were otherwise restricted.

11.   Argentina signed a nuclear cooperation agreement with India soon after the 1974 Indian nuclear explosion for research relating to plutonium and natural uranium. Neither country is a party to the NPT. The reactor of Canadian origin from which India obtained plutonium bears some resemblance to the Atucha I reactor of German technology.

12.   Pedro, Marta and Pouget, Mario G. Martin, op. cit.

13.   Brazil acquired a uranium enrichment capability in 1987. It should be recalled that the fuel used by Brazil's reactors is enriched uranium,

despite Brazil's being the fifth largest world producer of yellowcake. Uranium concentrate was sent to Europe for enrichment and then sent back to Brazil for use in reactors. This is one of the main reasons that the Geisel Government, near the end of its term in office (March, 1979), decided to develop technology that would provide full control of the fuel cycle; this decision gave rise to the so-called autonomous or parallel plan.

14. The Brazil CNEN Chairman, Rex Nazareth, defended the decision to promote the parallel program, stating: 'to allow autonomous and independent development that suits society's needs and not controlled by what other countries consider Brazil should be allowed to do. We have not foresworn technological independence, nor will we be a colony because of lack of technology.' *Folha de São Paulo*, March 20, 1987.

15. Neither Brazil nor Argentina adheres to NPT regulations. In their international speech they stated that the NPT 'disarms those without arms' and that its system of regulations hides hegemonic designs and monopolistic economic interests behind the legitimate principle of horizontal proliferation. Regarding the Treaty of Tlatelolco, it should be recalled that it was not conceived by the Latin American players for use as an instrument of pressure on the nuclear powers. Recently the Argentinean Foreign Minister Domingo Cavallo, referring to journalistic reports of a probable change in Argentina's position, stated that there were no changes in the traditional position, but acknowledged that 'there could be change if we Latin Americans found ways of creating new safeguard mechanisms administered by ourselves.'

16. On the nuclear policy debate in Brazil – in contrast to Argentina – see: Machado, Antonio, Candotti, Ennio *et al.*, Energia Nuclear e Sociedade, Rio de Janeiro, Paz e Terra, 1980; Gabeira, Fernando, Greenpace, verde guerilha de pas, São Paulo, Ed. Clube do Livro, 1988; Fullgraf, Frederico A., Bomba Pacífica. O Brasil e a corrida nuclear, São Paulo, Ed. brasiliense, 1988.

17. Martinez-Vidal, Carlos and Ornstein, Roberto, *La cooperacion argentino-brasilena en el campo de los usos pacificos de la energia nuclear*, in Hirst, Monica (comp.), Argentina-Brasil: Perspectivas comparativas y Ejes de Integracion, Buenos Aires, Ed. Legasa (in press).

18. Ibidem.

19. Brasil's CNEN Chairman, Rex Nazareth, clarified the situation regarding openness by saying: 'we are not going to tell what they do not tell us', *Folha de São Paulo*, April 6, 1989.

20. OPANAL (Organization for the Prohibition of Nuclear Arms in Latin America) congratulated the Presidents of Brazil and Argentina for the joint communiques – especially the one of July 30, 1986, which reaffirms their commitment to the peaceful use of nuclear energy, see Excelsior, Mexico, 21 August 1986.

21. On the subject of military cooperation, see Cagnari, Geraldo Lesbat,

Autonomia Estratgica e Cooperacao Militar, in Hirst, Monica (comp.), Argentina-Brasil: perspectivas comparativas y ejes de integracion, in press. It should also be noted that today rivalry is expressed in the special programs (the Argentinean Condor II and the Brazilian SS-1000), without this implying that all the nuclear issues are closed, even though as both countries feel the pressures of the missile club (signatories to the Missile Technology Control Regime) it is to be expected that – as in the case of nuclear issues and the London Club – they would come together to resist pressures and to make progress in technological cooperation, such as resulted from the agreement recently signed during President Menem's visit to Brazil (August, 1989).

22. See statements of Admiral (RE) Carlos Castro Madero, former Chairman of the CNEA during the last military government, in *La Ciudad de Belgrano*, Ano I, Nro. 14, June 1989, University of Belgrano, Buenos Aires.

23. Russell, Roberto, *La Posicion de Argentina frente al desarme, la no proliferacion y el uso pacifico de la energia nuclear*, in Desarme y Desarrollo *et al., Buenos Aires, GEL-Fundacion Illia, 1989.*

24. Hirst, Monica (in collaboration with Hector Eduardo Bocco), *La posicion brasiliera en materia de desarme y proliferacion nuclear*, in Desarme y Desarrollo *et al.*, op. cit.

25. This attitude was denounced in the first meeting of member countries of the Peace and Cooperation Zone that took place in Rio de Janeiro in 1987.

26. This concern has been expressed especially by the Brazilian Navy through Admiral Mario Cesar Flores, Chief of Staff of the Brazilian Navy.

27. Brazil was reacting to the proposal to eliminate 20% of the foreign debt through investment in an international organization for protection of the Amazon region, suggested at the VI Ministerial Meeting on the Environment of the Countries of Latin America and the Caribbean, organized by the United Nations Programme for the Environment. See *Gazeta Mercantil*, Rio de Janeiro, March 3, 1989. Brazil's position with respect to an attempt to link environmental protection with nuclear proliferation, shared by Argentina, was expressed by Ambassador Sergio Arruda in developing the thesis of *sustained development*. Brazil's Ambassador to the United States, Paulo Flecha de Lima, described the situation as 'the hardest pressures that Brazil has experienced in the course of its history.'

28. See interview of Josef Goldblat, member of SIPRI, in *Cono Sur*, Santiago de Chile, FLACSO, Vol. VII, Nro. 3, May–June 1988.

29. The outward development model to explore in the Argentinean case offers business enterprises such as INVAP (State company that groups the CNEA with the government of the Province of Rio Negro) which, in the context of economic crisis, energy crisis and financial restrictions, has demonstrated competitiveness in foreign markets and

inventive ability in the most diverse fields of technology (from agro to air-and-space projects) and a praiseworthy production capacity. The *Nur* reactor (installed in Algeria) and the CAREM project (the only *mini* reactor in supply on the international market) depend on this business enterprise.

30. See *Excelsior*, Mexico, August 21, 1986.
31. On this subject see Rizzo de Oliveira, Eliezer *et al.*, As Foras Armadas no Brasil, Espao e Tempo, Rio de Janeiro, 1987. Also Quartim de Moraes, Joao, *A tutela militar no Brasil da transiao controlada democracia blogueada*, in Hirst, Monica (comp.), Argentina-Brasil: Perspectivas . . . . op. cit.
32. Buenos Aires (1987), São Paulo (1988), Buenos Aires (1989).
33. See Bocco, Hector E., Atlantico Sur: Zona de Paz y Cooperacion, Masters Thesis, Flasco, 1988.
34. See Grandi, Jorge, *La Integracion, la cooperacion argentino-brasilera y la disuasion nuclear desarmada*, in America Latina-Internacional, Vol. 3, Nro. 10, October–December 1986, Flasco, Buenos Aires.
35. See Subramaniam, R. R., and Subrahmanyam, K., *Mutual Inspection and Verification*, in Subrahmanyam, K. (comp.), India and the Nuclear Challenge, Institute for Defence and Analysis, New Delhi, 1986.
36. Carasales, Julio C., El desarme de los desarmados: Argentina y el Tratado de No Proliferacion de Armas Nucleares, Buenos Aires, Ed. Pleamar, 1987.
37. In restructuring and unifying the Brazilian nuclear program, the Navy's Aramar Plant was not affected by the eight decrees signed by President Sarney.

Index